Fundamentals of Optimization

Mark French

Fundamentals
of Optimization

Methods, Minimum Principles,
and Applications for Making Things Better

 Springer

Mark French
School of Engineering Technology
Purdue University
West Lafayette, IN, USA

ISBN 978-3-030-09426-3   ISBN 978-3-319-76192-3   (eBook)
https://doi.org/10.1007/978-3-319-76192-3

This Springer imprint is published by the registered company Springer International Publishing AG part of Springer Nature.
The registered company address is: Gewerbestrasse 11, 6330 Cham, Switzerland

Foreword

I have had the great pleasure in getting to know Mark French's dedication and innovative ways of teaching engineering principles over the last few years. Mark is truly an exceptional teacher and this is reflected in his latest work: *Fundamentals of Optimization: Methods, Minimum Principles, and Applications for Making Things Better.* The best textbook authors for education tend to be those that have a strong desire to simplify and present difficult topics in such a way as to carefully lead the student along a learning journey that builds self confidence in the subject. Being an author of numerous textbooks myself, I know that dedicated teachers make great textbook authors and Mark's dedication is reflected in his book.

This book is an introductory treatise on the topic of optimization in engineering design. The point of this book is to provide a reference for students and professionals who need a clear introduction to optimization, with actionable information and examples that show how the calculations work. This is basically the book Mark French needed when he learned to use optimization as a civilian aerospace engineer with the US Air Force.

There are a number of mathematically rigorous optimization books in print and some are very good, but they are generally geared towards researchers and advanced graduate students. They tend toward abstract and theoretical presentations, often going so far as to introduce subjects using a mathematically formal statement-proof structure. While this is quite appropriate for the research community, it is the wrong approach for students who need to learn how to improve processes and products.

The main features of the book that make it useful for teaching and learning include the following:

- A focus on meaningful, less abstract problems
- Sample calculations including intermediate results
- Preference for accessible, robust methods
- Examples showing implementation in MATLAB
- Graphical results whenever possible

- Historical and industrial examples to give context to the development of optimization

Each chapter steps the reader through example problems, which include text explaining the process, the calculations, and the graphic representation of the results. This step-by-step learning process is an ideal teaching methodology that leads the learner through the concept being presented in a format that is easy to follow and understand. The use of real-world examples is especially compelling as it has the potential to draw the learner into the problem. As we know a self-motivated learner tends to be more engaged and interested in the topic.

Overall, Mark French has written a book that is very compelling in its approach, and his writing style and use of examples are some of the best I have seen in mathematics textbooks. Mark is an outstanding classroom teacher and that is reflected in this book.

Purdue University, West Lafayette,   Gary Bertoline
IN, USA

Preface

This is an optimization book for people who need to solve problems. It is written for those who know very little – maybe nothing – about optimization, but need to understand basic methods well enough to apply them to practical problems. A survey of available books on optimization finds a number of very good ones, but often ones written using high-level mathematics and sometimes even written in a statement-proof format.

Clearly, these books are valuable for people wishing to work at a high level of technical sophistication, but they are sometimes opaque to people needing, rather, an accessible introduction. I think there needs to be a book that brings a reader from a point of knowing next to nothing to a point of knowing the big ideas, the necessary vocabulary, a group of simple and robust methods, and, above all, a sense of how optimization can be used.

This book has resulted from years of teaching optimization to both engineering and engineering technology students and from years spent applying optimization methods to problems of practical interest. Its content has been tested on many classrooms full of inquisitive students, and only the lessons they think were successful appear here.

The number of methods presented here is small; I have selected a few simple, robust methods to describe in detail so that readers can develop a useful mental framework for understanding how information is processed by optimization software. While more sophisticated methods are certainly more efficient, they don't look that different to the user. Thus, this book sticks to the simple, robust, and accessible methods.

I have used two calculation tools, Mathcad and MATLAB. MATLAB is, at this writing, probably the standard tool for general technical calculation, at least in the United States and much of Europe. The more sophisticated plots and more extensive calculations in this book were generally done using MATLAB. However, MATLAB is a programming language that looks familiar to anyone who ever learned FORTRAN. While quite powerful, MATLAB code, especially efficient, vectorized code, can be very difficult to read – a real problem for a book like this.

Mathcad is a very different animal, designed to make smaller calculations simply and with little need for formal programming. It is essentially a math scratchpad that is easy to use and easy to read. I chose to illustrate the details of specific calculations using Mathcad because even readers who are new to Mathcad can probably read the sample files, even if they could not write them. The Mathcad files in this book were made using Mathcad V15. The name Mathcad is not, in my opinion, a very good one, though I very much like the software. It has nothing to do with CAD or graphics software.

Finally, I hope this book will serve the needs of analysts for whom the result is more important than the algorithm. Those who wish eventually to study the behavior of algorithms and to perhaps even develop better methods may, I hope, find this volume a useful start before moving on to more advanced texts.

West Lafayette, IN, USA   Mark French

Acknowledgments

Every author knows that a book is the result of many contributions. My students continue to make sure that I know what I'm talking about and remind me when my explanations aren't clear enough.

When I was a new engineer at what was then called the Air Force's Flight Dynamics Lab, Ray Kolonay, Bob Canfield, and V.B. 'Van' Venkayya collectively helped teach me how the ideas of optimization worked and how to apply them. Working on their ASTROS project was a time of great learning for me and I hope I was of some value to them.

Over the years, I have been fortunate to participate in many different projects and work with some outstanding colleagues. From them, I learned about how the engineering concepts I learned in school are applied. These people are too numerous to name here and I hope they will forgive the omission. Perhaps it is enough to simply acknowledge my debt to them.

Kay Solomon continues to be the person to whom I can go for help with line editing and for help in turning clumsy phrasing into something more readable. Her literary fabulosity glows undiminished. Everyone should, as I do, know someone who likes to read the dictionary.

I am fortunate to work in a place where I can do things like write optimization books. I'm not a conventional professor and would probably be an interference fit in most academic environments. It is my great good fortune to have found a place where I can thrive.

My beloved wife, Amy, has now suffered through three books and, inexplicably, retains her patience and good humor.

Finally, I wish to thank Ken Burbank for being the best boss I've ever had. This book is dedicated to him.

Contents

Chapter 1
Optimization: The Big Idea

Continuous improvement is better than delayed perfection

-Mark Twain

We all naturally know something about optimization. Whenever you think that an object or a process could be made better, you are thinking about optimization. It could be something as simple as deciding whether going straight or turning at a light gets you home faster. It might be something more complicated like trying to find a way to make a deck with little wood as possible.

The results of optimization are all around us. There is a 355 mL (12 oz) can of diet soda sitting on the table next to me right now. The aluminum soda can may be one of the most optimized products many of us see. Hundreds of billions of them are made every year. Reducing the cost of each can by even a small fraction of a cent could result in big savings when added up over a few billion cans.

The lowly drink can is designed to be inexpensive, hold a certain volume of liquid, be easy to open, be sanitary, be easy to manufacture in very large quantities, and be strong enough not to fail when stacked in a warehouse or in a hot truck driving down a bumpy road. Over time, its design has evolved so that the result is a highly optimized product.

Decades ago, drink cans were made of steel and had no means of opening built in. Rather, they were opened using a small tool often called a churchkey that people had to keep available. Figure 1.1 shows such a can with the triangular openings made by a churchkey. People usually made two openings so air could enter the can to replace the liquid. Note also that the can is truly a cylinder, there is no taper at the top as in modern cans.

In the latter part of the twentieth century, aluminum cans replaced steel ones, and pull tabs were introduced so that churchkeys were no longer needed. The original pull tab separated from the can when it was opened and was usually discarded (Fig. 1.2). Pull tabs became, perhaps inevitably, a source of materials for many craft projects. However, they were also discarded in large numbers and were perceived by many people as a nuisance. As a result, can designs were changed so that the pull tabs remained attached to the can [1].

© Springer International Publishing AG, part of Springer Nature 2018
M. French, *Fundamentals of Optimization*,
https://doi.org/10.1007/978-3-319-76192-3_1

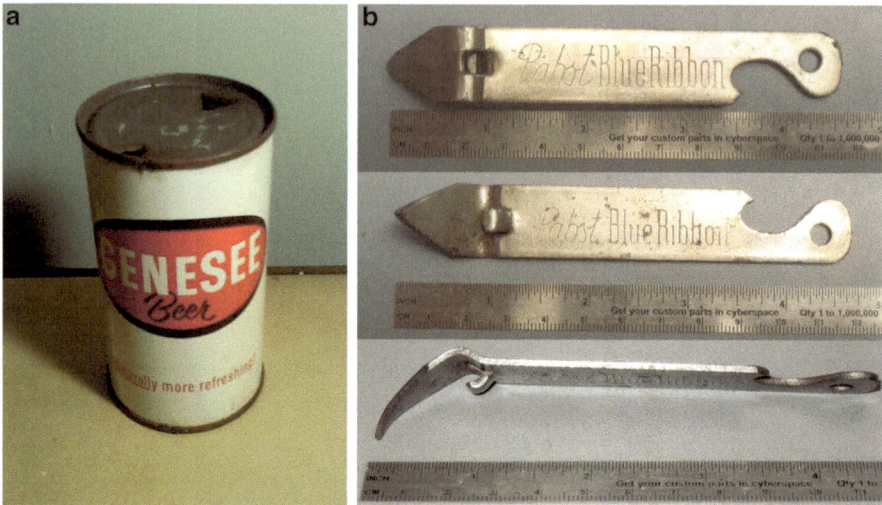

Fig. 1.1 An old beer can and the type of churchkey that may have been used to open it. (Wikimedia Commons, image is in the public domain)

Fig. 1.2 An old pull tab. (Wikimedia Commons, image is in the public domain)

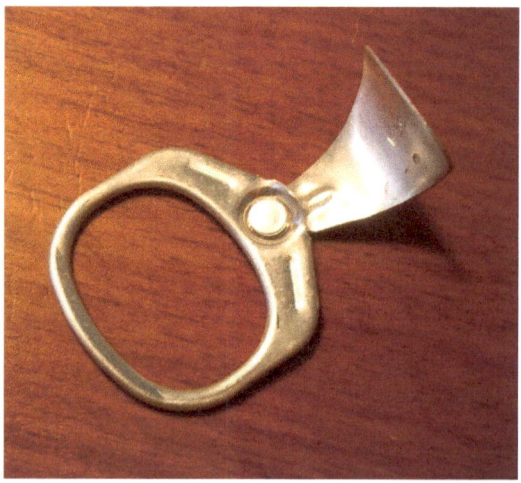

The modern pop can (Fig. 1.3) is, thus, the result of a long chain of design improvements. The walls are very thin, but it is strong enough to withstand loads from being stored and transported. It is highly optimized, but there may be room for further improvements [2, 3]. Future designs may be cheaper while still satisfying all the constraints on the design.

Once you learn the basic ideas of optimization, you will start to see optimization everywhere. For example, anyone familiar with auto racing already knows something about optimization. I'm writing this in the summer and the Indianapolis 500 is on TV. There is hardly a clearer example of optimization than this race. The goal is

Fig. 1.3 A modern pop can.
(Photo by the author)

obvious – to run 200 laps (500 miles) and be the first to cross the finish line. There are many different ways to do this.

The number of choices a team can make along the way is huge. For example, some paths around the track allow higher average speeds than others. The driver can obviously choose how fast to drive. Running too fast early in the race may cause the car to run out of fuel before the end of the race, but running too slowly will leave the car so far behind the rest of the pack that it cannot catch up. Changing settings on the car, like wing angles, may make it faster, but at the cost of higher fuel consumption and faster tire wear, which may, in turn, require extra pit stops. The list of possible choices is a very long one.

While most of the choices made by the race teams may be transparent to the average fan, one is very clear – the path taken by drivers around the track. There is one particular path that is faster than all the others, often called the groove [4, 5]. The groove is easy to see on a race track. Since it's where drivers spend most of their time, it's where they leave the most rubber. The resulting groove is a dark stripe around the track. Figure 1.4 shows the groove around one of the turns at the Indianapolis Motor Speedway.

Some of the ways of increasing speed around the track are subtle and nonintuitive. The announcer just made the comment that two cars running together are always faster than one. When one car closely follows another one, the aerodynamic drag of the pair is lower than the drag of the two cars separately. Thus two teammates can cooperate so that they both either go faster or burn less fuel. A more delicate situation occurs when competing drivers find themselves in a situation where the same kind of cooperation might benefit them both.

Fig. 1.4 A turn at the Indianapolis Motor Speedway. (Wikimedia Commons, image is in the public domain)

The cars themselves are very optimized for their specific task. They can go about 80 miles (129 km) on a tank of gas and need to change tires about as often as they refuel. They cannot drive in even a light rain because of the danger of hydroplaning, and they can accommodate only relatively small drivers. In return for these limitations, though, the result is cars than drive at high speed for a long distance on a closed course [6].

The concept of optimization is probably an intuitive one for most of us; we are used to the idea of looking for ways to make things better. However, applying optimization methods to real problems requires a way to describe those problems precisely.

The first step along the path is to clearly define three specific ideas: the objective function, design variables, and constraints.

The objective function is the quantity to be maximized or minimized. For the pop can, the objective function is cost. For the race driver, the objective function is time to complete the race. Design variables are the quantities that can be changed in order to change the objective function. Finally, constraints are limits placed on the value of design variables. All these quantities live in a mathematical place called design space [7].

1.1   Design Space

Design space is multidimensional, with one dimension for each design variable and an additional one for the objective function. Solving an optimization problem simply means finding the point in design space that corresponds to the minimum value of the objective while satisfying the constraints. If design space has only two or three dimensions, that is, one or two design variables, then it is usually possible to simply make a plot of design space.

Optimization is often presented in abstract, mathematical terms, but it doesn't need to be that way, at least for simple problems. As humans, we have evolved to look for patterns and to think in terms of pictures. Indeed, mathematics can be defined as the study of patterns.

Before Rene Descartes, algebra and trigonometry were different subjects. He changed that by introducing what we now call Cartesian coordinates with both horizontal and vertical axes. Now, we teach children to plot functions with the independent variable mapped to the horizontal axis and the dependent variable on the vertical axis. Cartesian coordinates turned algebra into pictures.

For example, the equation $y = x^2$ describes a parabola. This simple equation has been the subject of study since the ancient Greeks. However, it wasn't always written out that way. They thought of parabolas as sections cut from cones as shown in Fig. 1.5.

Rene Descartes unified geometry and algebra by making it possible to draw pictures of mathematical expressions. Because of this, Newton could use his calculus to show that the path of a body in flight was a parabola. The water jets spraying from the nozzles in Fig. 1.6 follow parabolas just as Newton predicted.

Just as we can define an optimization problem mathematically, we can draw a picture of it if there are only one or two design variables. The framework for that

Fig. 1.5 Conic sections. (Wikimedia Commons, image is in the public domain)

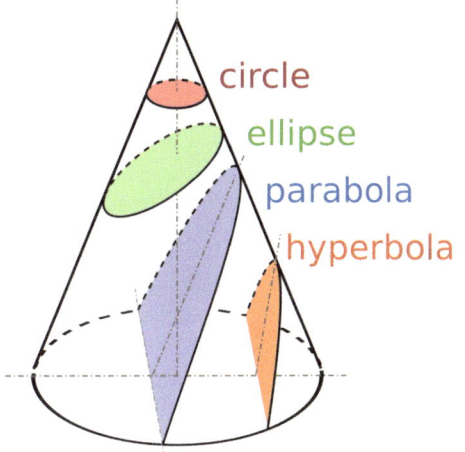

Fig. 1.6 Water jets forming
parabolas. (Wikimedia
Commons, image is in the
public domain)

picture is called design space. In design space, the independent variables and the
dependent variable are all mutually perpendicular.

Imagine the simplest possible example, where there is one independent variable
and one dependent variable. Here's the problem statement:

> Given a fixed launch velocity and assuming that a ball travels in a parabolic path, find the
> launch angle that gives the longest range.

This simple statement defines the objective function – range – and the single design
variable – launch angle. We know already that design space will have two dimensions, one describing the launch angle and one describing the range. The problem
now is how to mathematically relate the two.

From basic dynamics, the description of the path of the ball is

$$y = -\frac{g}{2}\frac{x^2}{v_0^2 \cos^2\theta} + x\tan\theta \tag{1.1}$$

where g is the acceleration of gravity (9.81 m/s^2) and v_0 is the initial velocity. The
goal is to find a launch angle, θ, which gives the maximum range for a given initial
velocity. When the ball hits the ground, $y = 0$.

Let's assume that the launch point is at the origin of the coordinate system, so the
ball impacts when $y = 0$. We can find the range, x, by solving Eq. (1.2).

$f(\theta) := \sin(\theta) \cdot \cos(\theta)$ <= Define the objective function

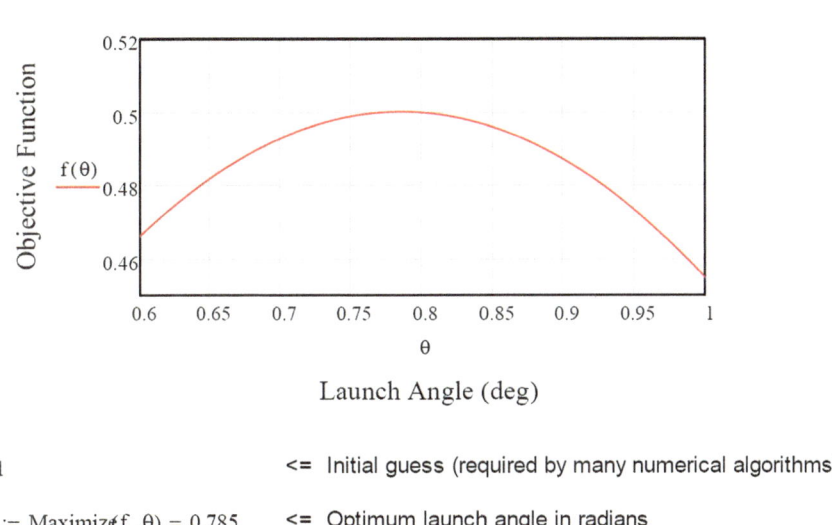

Launch Angle (deg)

$\theta := 1$ <= Initial guess (required by many numerical algorithms)

$\theta_{max} := \text{Maximize}(f, \theta) = 0.785$ <= Optimum launch angle in radians

$\theta_{max} = 45 \cdot \text{deg}$ <= Optimum launch angle in degrees

Fig. 1.7 Mathcad solution of the maximum range problem

$$0 = -\frac{g}{2} \frac{x^2}{v_0^2 \cos^2 \theta} + x \tan \theta \qquad (1.2)$$

Solving for x using the quadratic equation gives

$$x = \frac{2v_0^2}{g} \cos^2 \theta \tan \theta \qquad (1.3)$$

which simplifies to

$$x = C \sin \theta \cos \theta \qquad (1.4)$$

where $C = 2v_0^2/g$. Since C is a constant, we can divide though by C to get

$$\frac{x}{C} = f(\theta) = \sin \theta \cos \theta \qquad (1.5)$$

The axes on which this function is plotted form the design space. The independent variable is θ, and the dependent variable is f where $f = x/c$. Finding the value of θ that maximizes f is easy to do using a calculation package like Mathcad as shown in Fig. 1.7. It is important to note that the curve in this figure is not a parabola and does not show the trajectory of the ball through space. Rather, it shows the range as a function of launch angle. Any resemblance to a parabola is coincidental.

Note that this result doesn't require knowing the launch velocity. Within the limits of our analysis (no aerodynamic resistance), the maximum range occurs at a launch angle of 45°, no matter what the launch velocity is.

It is, of course, possible to solve this problem analytically if we like since we have a simple expression for the objective function. We can find the derivative, df/dx, set it equal to zero, and solve for θ. If you would like a brief reminder of how derivatives work, see Appendix A.

$$\frac{df}{d\theta} = \frac{d}{d\theta} \sin \theta \cos \theta = \cos 2\theta = 0 \qquad (1.6)$$

$\cos(90°) = 0$, so $2\theta = 90°$ and $\theta = 45°$.

The "maximize" function in Mathcad does the equivalent of setting df/dx equal to zero and also ensuring that the curvature is negative. However you wish to look at the problem, we have solved an optimization problem of one design variable.

1.2   What Is Optimum?

Before moving on to more involved problems, it makes sense to pause and think about what it means to have an optimum design. Without a good answer to this question, it is impossible to write out an optimization problem in mathematical terms. In optimization problems, all calculations follow from the decision of what quantity is to be maximized or minimized and what constraints need to be satisfied.

Imagine you are given the task of designing a car. An optimal car might be especially efficient or light or safe or have any number of other desirable traits. However, in order to produce an optimum design, you have to pick something as an objective function. Some methods allow more than one objective function, but having too many objective functions may eventually become the same as having no objective function. As an example, it's easy to think that we might want the safest car possible and, indeed, attempts have been made to design cars with safety as the primary consideration. However, this hasn't always been a successful approach.

It turns out that safety features, when taken to the extreme, are heavy and expensive. The result has generally been safe cars that aren't very good cars. A notable example was the Bricklin SV-1, introduced in 1974. It was designed to be a safe and affordable sports car. However, the weight of all the safety features so limited the performance that it was unattractive as a sports car. It was only in production for 3 years, and it now routinely appears on lists of the worst cars of all time [8]. It was safe by the standards of the time, though (Fig. 1.8).

As an even more extreme example, consider a design in which acceleration has taken priority over all else, a top fuel dragster. It is a vehicle optimized to accelerate from a standing start to 305 m (1000 ft) in as short a time as possible. The current record is about 3.7 s and the current top speed is about 530 km/h (330 miles/h).

Fig. 1.8 Bricklin SV-1 safety sports car. (Wikimedia Commons, image is in the public domain)

The top fuel dragster is so optimized that it is essentially useless for anything other than its intended purpose. Figure 1.9 shows a representative car on the track.

A complete list of the characteristics that have resulted from the continuous refinement of the designs is not possible here, but a few items [9, 10] might be enough to make the point:

- Time to reach 160 km/h (100 miles/h) is about 0.8 s, and time to reach 322 km/h (200 miles/h) is about 2.2 s.
- At full power, the engine makes about 6000 kW (8000 hp), though this has not been measured directly, because no engine dyno can absorb that much power that quickly. This value is calculated and estimates range as high as 7500 kW (10,000 hp).
- There are two spark plugs per cylinder and two magnetos, each generating about 22 amps. The plug electrodes are completely consumed during a pass. After the midpoint, the engine is dieseling from compression and hot exhaust valves.
- The service life of the engine at power (one burnout and one run) is a little less than 1000 crankshaft revolutions.
- The vehicle can reach 480 km/h (300 miles/h) in about 550 crankshaft revolutions.
- The engine burns nitromethane. It requires less oxygen than gasoline, so more of it can be put into the engine (less room needed for air), and it can generate about 2.3 times as much power.

Fig. 1.9 A top fuel dragster. (Wikimedia Commons, image is in the public domain)

- The supercharger requires about 450 kW (600 hp) to turn. A stock Dodge Hemi V-8 engine could not even power the supercharger on a top fuel car.
- Average acceleration is more than 4 g. Instantaneous deceleration on deployment of the braking parachute is about 6 g.
- The connecting rods plastically deform during each run, so the compression ratio at the end of the run is different than at the beginning of the run.
- The engine must be rebuilt after every run.

Top fuel dragster design is limited by many rules (constraints) designed to limit top speeds and to ensure an acceptable level of safety. For example, rules do not allow the engine to respond to real-time information, so the ignition system cannot be computer-controlled. Another constraint designed to limit the top speeds is that the final drive ratio cannot be more than 3.2:1. For safety, the driver must be protected by an enclosed roll cage.

The resulting vehicle is beautifully optimized, with much of it operating at the limits of what is physically possible. However, it can't do anything else than run on a drag strip. Highly optimized products or processes can become this specialized.

Top fuel dragsters are extremely cool, but not at all practical. Clearly, designing a practical car must require more than simply selecting an objective function. Rather than some highly optimized race car, let's consider a very popular small sedan, the Honda Civic shown in Fig. 1.10. It has sold very well over its long life because it has a very attractive combination of characteristics. It is inexpensive to buy and to operate, it is safe, it is comfortable, it is durable, and it is attractive to buyers.

Fig. 1.10   A Honda Civic Sedan. (Wikimedia Commons, image is in the public domain)

So, think about the ubiquitous Honda Civic, and think about what the design team might have selected as an objective function. Unlike a race car that only has to do one thing well, the Civic has many possible uses, and the resulting design had to be a compromise between many conflicting requirements. The point is that, for some designs, it is not always possible to identify a single objective function. In response to problems like this, a class of optimization methods has emerged that allows more than one objective function.

It is, of course, possible to select requirements that are so opposed to one another that an acceptable design just isn't physically allowable. A clear example is that of the flying car [11]. The idea has been that cars and airplanes are both very useful in their respective ways. Thus, it must be even more useful to develop a machine that can act as both a car and an airplane. An early example was the Convair 118 as shown in Fig. 1.11.

The vehicle used a 19 kW (25 hp) air-cooled engine for operation on the road and a 140 kW (190 hp) for flight. It first flew in 1947 and did, indeed, function as both a car and an airplane. It was apparently not a very good car or a very good airplane because plans for production were cancelled after initial testing.

The lure of the flying car (or roadable aircraft, take your pick) is such that several more have been developed since the Convair 118. To date, none have entered production, in spite of determined efforts. So far, everyone has pretty much come to the same conclusion: making a machine that can both fly and drive results in a bad car that is also a bad airplane.

Taking another approach, let's consider durability. This might not make a very good objective function since it would result in a very heavy and expensive car (think armored military vehicle), but it is routinely used as a constraint. In the USA, manufacturers often test their designs to 150,000 miles (241,000 km) of simulated use. The goal is to make vehicles that last long enough to satisfy buyer expectations without making them so heavy or expensive as to be unattractive.

Fig. 1.11 Convair 118 flying car. (Wikimedia Commons, image is in the public domain)

Let's say it needs to last 10 years. Most buyers would be happy with a car that lasted through 10 years of daily use. What if some part, say the transmission, fails after 10 years and 1 day? At first, that might sound acceptable, but what about the rest of the car? The rest of the car is still in working order.

The components of the car that are still working after 10 years are apparently overbuilt. That suggests they are too durable and, thus, maybe too expensive or too heavy. If that's the case, then maybe the rest of the car should be designed so that more of the components will fail after about 10 years. Perhaps all the components of the car should fail at 10 years and 1 day.

If every part failed at the same time, then there would be no excess service life in the car, and it would presumably be as inexpensive as it could reasonably be. Imagine a car that needed nothing more that minor service for 10 years and then just went pieces in the driveway. In a very real sense, that would be an optimum design.

This idea has been around for a while now. In 1858, Oliver Wendell Holmes wrote a light poem called "The Deacon's Masterpiece or the Wonderful One-Hoss Shay." A one-horse shay is a type of light buggy pulled by a single horse (Fig. 1.12). Holmes describes a buggy that works perfectly until the first part fails. Then, the entire buggy falls to pieces at once. Clearly, people have been thinking about optimal design for a long time.

Fig. 1.12 A one-horse shay. (Digital Commonwealth, image is in the public domain)

References

1. Rogers LF, Igini JP (1975) Beverage can pull-tabs- inadvertent ingestion or aspiration. J Am Med Assoc 233(4):345–348
2. Yamazaki K, Itoh R, Han J, Nishiyama S (2007) Applications of structural optimization techniques in light weighting of aluminum beverage can ends. J Food Eng 81(2):341–346
3. Han J, Itoh R, Nishiyama S, Yamazaki K (2005) Application of structure optimization technique to aluminum beverage bottle design. Struct Multidiscipl Optim 29(4):304–311
4. Timings JP (2012) Vehicle trajectory linearisation to enable efficient. Vehicle Syst Dyn 50 (6):883–901
5. Velenis E, Tsiotras P (2008) Minimum-time travel for a vehicle with acceleration. J Optim Theory Appl 138(2):275–296
6. Seward D (2014) Race car design. Palgrave – MacMillan Education, London
7. Vanderplaats G (2001) Numerical optimization techniques for engineering design, 3rd edn. Vanderplaats Research and Development, Colorado Springs
8. Neil D. 50 worst cars of all time. Time [Online]. Available: http://content.time.com/time/specials/2007/article/0,28804,1658545_1658533,00.html. Accessed 12 June 2016
9. NHRA. NHRA drag racing 411. [Online]. Available: http://www.nhra.com/streetlegal/nhra101.aspx. Accessed 12 June 2016
10. Top fuel by the numbers. Motor Trend [Online]. Available: http://www.motortrend.com/news/top-fuel-numbers/. Accessed 6 June 2016
11. Drive right up. Popular Science, p 92, April 1946

Chapter 2
Getting Started in Optimization: Problems of a Single Variable

If you can analyze, you can optimize

-Ray Kolonay

The simplest optimization problems have a single design variable and no constraints. A surprising number of useful problems can be described using only one design variable, and they can be a good way of learning how to solve optimization problems.

A good way to see how optimization problems are formulated is to study examples that address practical problems. What follows are some examples of useful single variable, unconstrained optimization problems. To start, they are solved by simply plotting the objective function and looking for the minimum. Then, we'll see some solution methods that can solve single variable problems using general algorithms. First, let's look at a simple example of getting from one point to another in the shortest possible time.

2.1 The Lifeguard Problem

Most of us have been to a beach and seen lifeguards watching for swimmers in trouble. Lifeguards are usually placed at intervals along the beach (Fig. 2.1). If a lifeguard sees a swimmer in distress, he or she must run across the sand to the water's edge and swim out from there.

It's clear that the lifeguard can run faster across the sand than she can swim. Let's say that the lifeguard can run about 7 m/s across the sand and can cover about 2 m/s in the water. A good swimmer can go a little more than 1 m/s, but the lifeguard will be able to run through the shallow water before it gets deep enough that she needs to swim. It seems reasonable that the average speed through the water is about 2 m/s. This an approximate analysis, so these numbers are fine for now. We'll improve the model later.

Figure 2.2 shows the path from the lifeguard to the swimmer. The shortest distance between two points on a plane is a straight line, so we can assume the

M. French, *Fundamentals of Optimization*,
https://doi.org/10.1007/978-3-319-76192-3_2

Fig. 2.1 A lifeguard stand at a beach. (Wikimedia Commons – image is in the public domain)

lifeguard travels in a straight line across the sand and another straight line through the water.

The lifeguard wishes to reach the swimmer in the shortest possible time (and the swimmer probably agrees). A straight line from the starting point to the swimmer is the shortest distance. However, since the speed over sand and the speed through water are different, it is not the path that minimizes time.

The problem, then, is to find a path to the swimmer that minimizes time. The objective function is time, and the design variable is x, the point at which the lifeguard enters the water. To turn this problem statement into an optimization problem, the objective function must be written as a function of the design variable.

We all learned as children that time $\times$ rate = distance assuming the rate, or velocity, is constant. It follows then that time = distance/rate. Also, the time required to get to the swimmer is the time over the sand plus the time over the water.

$$T = T_s + T_w = \frac{L_s}{V_s} + \frac{L_w}{V_w} \tag{2.1}$$

It is simple now to define the distances as functions of the design variable using the Pythagorean theorem.

$$L_s = \sqrt{50^2 + x^2}$$
$$L_w = \sqrt{50^2 + (100 - x)^2} \tag{2.2}$$

So, the objective function is

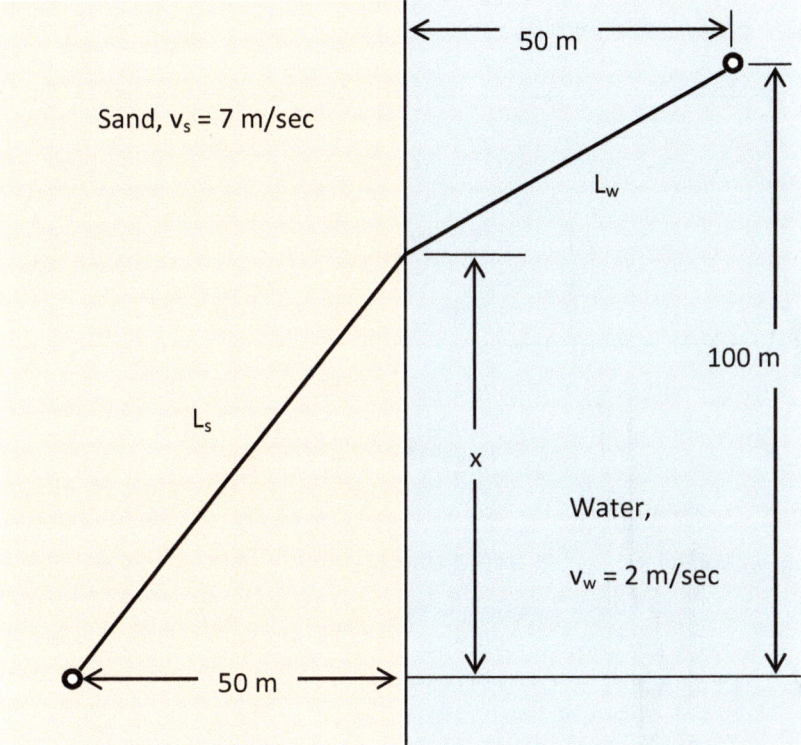

Fig. 2.2 Layout of the lifeguard problem

$$T(x) = \frac{1}{7}\sqrt{50^2 + x^2} + \frac{1}{2}\sqrt{50^2 + (100 - x)^2}, \quad (2.3)$$

and it is easy to find the minimum time to the swimmer by plotting $T(x)$. Figure 2.3 shows one way of finding the solution. This one is done using Mathcad.

It is traditional to label the solution with an asterisk [1], so the value of x that gives the minimum time for the lifeguard to reach the swimmer is x^*. The minimum time to reach the swimmer is about 40.166 s, and, to achieve this, the lifeguard must run 87.208 m down the beach before getting into the water. The lifeguard can run much faster than she can swim, so it makes sense that she would run the largest portion of the way over the sand before entering the water.

There is one other thing to be learned from the plot. The minimum time occurs when the lifeguard runs 87.208 m up the beach. But, no lifeguard would pause to measure distance to three decimal points while running to save a swimmer (I hope). Rather, she is much more likely to simply estimate the correct distance and get on with it.

This approximate approach is quite acceptable since the objective function doesn't change very quickly in the neighborhood of the optimum, x^*. If the lifeguard

$$T(x) := \frac{1}{7}\cdot\sqrt{x^2 + 50^2} + \frac{1}{2}\cdot\sqrt{(100 - x)^2 + 50^2}$$

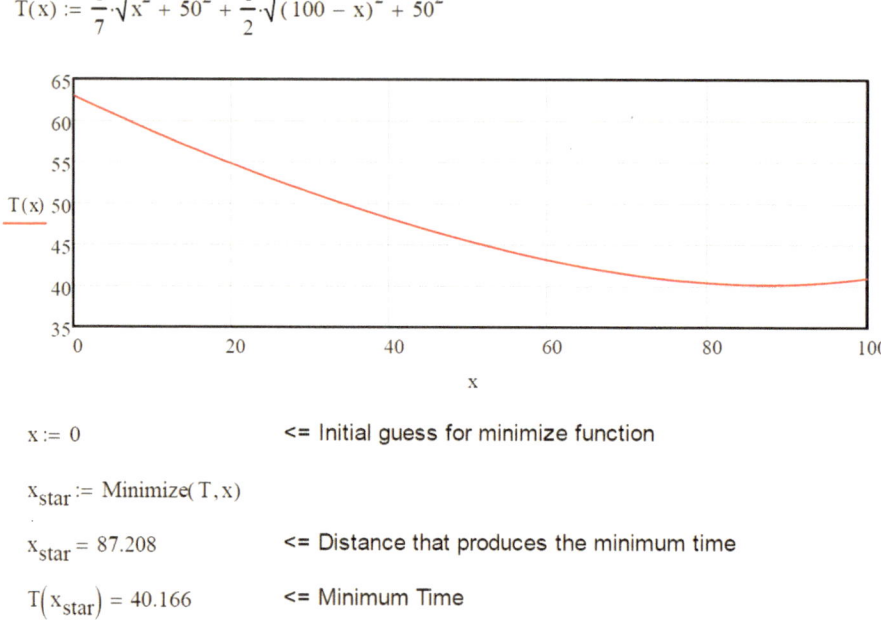

x := 0   <= Initial guess for minimize function

$x_{star} :=$ Minimize(T, x)

$x_{star} = 87.208$ <= Distance that produces the minimum time

$T(x_{star}) = 40.166$ <= Minimum Time

Fig. 2.3 Mathcad solution to the lifeguard problem

only ran 80 m up the beach, the time to reach the swimmer would be 40.4 s, and if she ran all the way up the beach, 100 m, the resulting time would be 41.0 s. This means that even an approximate solution to the problem will get pretty close to the minimum time – good news indeed for the swimmer.

In mathematical terms, the slope of the objective function at x^* is zero. Also, the curvature of the objective function (its second derivative) is low near x^*. That's why it doesn't change very much in the neighborhood of the optimum.

2.2   Maximum Range of a Projectile with Aerodynamics

The fact that there is only a single design variable doesn't mean that the underlying analysis needs to be primitive or incomplete. We showed earlier that the maximum range of a projectile occurs when the launch angle is 45°. However, we neglected the effect of aerodynamic drag. For a dense projectile that is reasonably streamlined and not traveling too fast, this is probably a good approximation. But, there are many projectiles for which this isn't true. Let's consider a tennis ball with a diameter of 65 mm and a mass of 57 g. Anyone who has ever put a spin on a tennis ball knows that aerodynamic forces can affect its path through the air, so we should expect a different result than when aerodynamics are ignored.

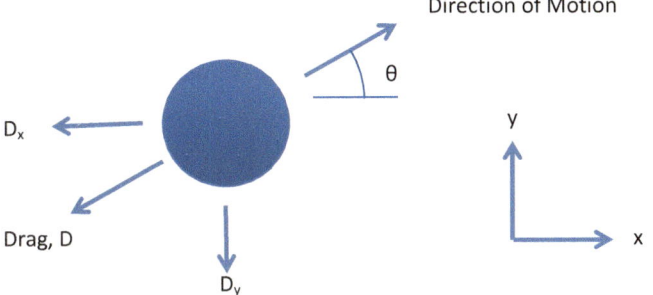

Fig. 2.4 Aerodynamic forces acting on a ball in flight

Since the ball can move in both horizontally and vertically, we will need to model its motion in two directions. We'll assume there is no spin, so the aerodynamic drag force just acts opposite to the direction of motion [2]. Figure 2.4 shows the ball with the components of aerodynamic drag.

The aerodynamic force, D, is defined as

$$D = \frac{1}{2} C_d \rho V^2 s \tag{2.4}$$

where:

C_d is a nondimensional drag coefficient, $C_d = 0.45$ for a sphere
ρ is air density, $\rho = 1.23$ kg/m^3 at sea level
V is velocity (m/s)
s is the cross-sectional area of the ball, $s = 0.010425$ m^2

Drag can be broken down into the x and y directions, so we can write Newton's law in both directions.

$$D_x = D \cos(\theta) \quad D_y = D \sin(\theta) \tag{2.5}$$

X direction $\sum F_x = m \, a_x$

$$-D_x = m \, a_x = m \, \ddot{x} \tag{2.6}$$

Y direction $\sum F_y = m \, a_y$

$$-D_y - mg = m \, a_y = m \, \ddot{y} \tag{2.7}$$

So the system of equations to be solved is

Fig. 2.5 Velocity
components

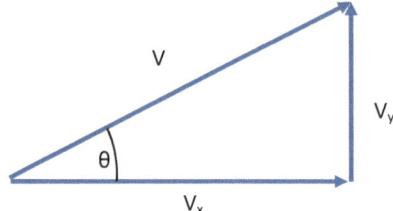

$$m\ddot{x} = -C_d \frac{1}{2}\rho V^2 s \cos(\theta) \tag{2.8}$$

$$m\ddot{y} = -C_d \frac{1}{2}\rho V^2 s \sin(\theta) - mg \tag{2.9}$$

We don't want θ to appear in the equations, so we will define θ in terms of velocity as shown in Fig. 2.5.

We can use the following relationships to put the two differential equations in their final form.

$$V = \sqrt{V_x^2 + V_y^2} \text{ and } V^2 = V_x^2 + V_y^2 \tag{2.10}$$

$$\cos\theta = \frac{V_x}{V} = \frac{V_x}{\sqrt{V_x^2 + V_y^2}} \text{ and } \sin\theta = \frac{V_y}{V} = \frac{V_y}{\sqrt{V_x^2 + V_y^2}} \tag{2.11}$$

Substituting these into the previous two equations gives

$$m\ddot{x} = m\dot{V}_x = -C_d \frac{1}{2}\rho s V_x \sqrt{V_x^2 + V_y^2} \tag{2.12}$$

$$m\ddot{y} = m\dot{V}_y = -C_d \frac{1}{2}\rho s V_y \sqrt{V_x^2 + V_y^2} - mg \tag{2.13}$$

Since aerodynamic forces are part of the problem, nothing conveniently cancels out like we saw when drag was neglected. We are left with a system of two ordinary differential equations, one for motion in the x direction and one for motion in the y direction. They can be written in terms of velocity, whose solution gives $V_x(t)$ and $V_y(t)$. Alternatively, they can be written in terms of position to give a pair of second-order differential equations, whose solution gives $x(t)$ and $y(t)$. In either case, the solution defines the path of the ball on the x-y plane.

Students are often conditioned to be wary of differential equations and perhaps you were as well. However, there is often no need for this because software exists that makes solving differential equations simple using robust numerical algorithms [3, 4].

There are two overarching ideas to remember when working with equations like these. The first is that a differential equation is just an equation with a slope in it. The second is that the solution to a differential equation is a function, not a number. If the equations are written using accelerations, the solutions will be two functions of time, $x(t)$ and $y(t)$.

Before we can solve these equations, we need some initial conditions. If we write the equations in terms of position, there are two second-order equations, so we will need four initial conditions. Let's assume the ball starts at the origin, so $x(0) = 0$ and $y(0) = 0$. Let's also assume that the initial velocity is 50 m/s and the launch angle is $45°$, so our initial conditions are

$$\begin{aligned}
x(0) &= 0 \\
y(0) &= 0 \\
\dot{x}(0) &= V_0 \cos \theta_0 \\
\dot{y}(0) &= V_0 \sin \theta_0
\end{aligned} \tag{2.14}$$

With all the pieces in place, it is simple to use Mathcad or some other program to find the solution. Since we have one equation describing motion in the x direction and one in the y direction, Mathcad has to find two equations, $x(t)$ and $y(t)$, that describe motion as a function of time. Figure 2.6 shows the solution. The resulting trajectory is clearly not a parabola, because of the effect of aerodynamics.

The solution in Fig. 2.6 shows the path for a single launch angle. In order to find the maximum range, we will need to find the range for a number of different launch angles and interpolate to find the maximum. A simple, robust way to do this is to fit a curve through the calculated ranges and find the maximum range using the curve as shown in Fig. 2.7.

The peak of this curve is at $\theta = 36.087°$ and predicts a maximum range of 76.19 m. Running the calculation again with the predicted optimum launch angle gives almost exact agreement with the curve fit.

2.3   Maximum Electrical Power Delivered to a Load

While the previous example is mechanical in nature, it is important to note that any object or process that can be described mathematically might be a candidate for optimization. Consider the problem of how to deliver the most power to an electrical load [5]. In electrical circuits, the impedances (dynamic equivalent of resistance) of the supply and of the load determine how much power can be transmitted. Think of an amplifier driving a speaker where the amplifier is the power source and the speaker is the load. It is common to treat the problem by modeling it as a DC circuit called a voltage divider. Clearly, a DC circuit is a very rough approximation of an AC amplifier and speaker, but the mathematical description is simple, and the result is approximately correct.

Figure 2.8 shows the circuit for a DC voltage divider, where:

V_{in} = input voltage delivered by amplifier
V_{out} = output voltage delivered to speaker

Ballistic Flight with Aerodynamic Drag

$C_d := 0.45$ $\rho := 1.23$ $D := 0.065$ $mass := 0.057$ $g := 9.81$ $S := \dfrac{\pi}{4} \cdot D^2$

$V_0 := 50$ $\theta_0 := 45 \cdot deg$ $T := 4.8$

Given $x''(t) = \dfrac{-1}{mass} \cdot C_d \cdot \dfrac{1}{2} \cdot \rho \cdot S \cdot x'(t) \cdot \sqrt{x'(t)^2 + y'(t)^2}$ $x(0) = 0$ $x'(0) = V_0 \cos(\theta_0)$

$y''(t) = \dfrac{-1}{mass} \cdot C_d \cdot \dfrac{1}{2} \cdot \rho \cdot S \cdot y'(t) \cdot \sqrt{x'(t)^2 + y'(t)^2} - g$ $y(0) = 0$ $y'(0) = V_0 \sin(\theta_0)$

$\begin{pmatrix} x \\ y \end{pmatrix} := Odesolve\left[\begin{pmatrix} x \\ y \end{pmatrix}, t, T \right]$ <= Ordinary differential equation solver

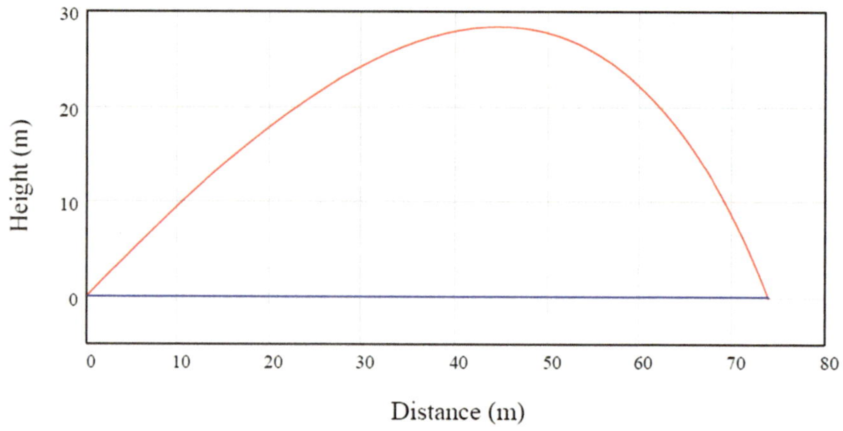

Fig. 2.6 Mathcad solution to ballistic motion with aerodynamics

$R_1 =$ output resistance of amplifier
$R_2 =$ input resistance of speaker

Note that, for historical reasons, when dealing with electrical circuits, i stands for current and not $\sqrt{-1}$. In circuit analysis, $j = \sqrt{-1}$.

Power delivered to the load (R_2) is $P_2 = i^2 R_2$ where i is the current flowing through the circuit. The two resistors are in series, so $V_{in} = i(R_1 + R_2)$ and $i = V_{in}/(R_1 + R_2)$. Substituting this into the expression for power gives

$$P_2 = \frac{R_2}{(R_1 + R_2)^2} V_{in}^2 \tag{2.15}$$

If the output resistance of the amplifier, R_1, is fixed, then there must be a value of R_2 that maximizes P_2.

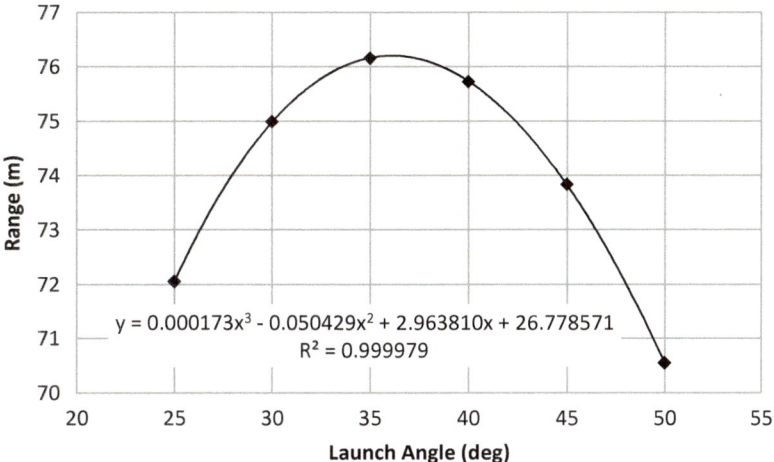

Fig. 2.7 Interpolation to find maximum range

Fig. 2.8 Voltage divider circuit

V_{in} = Input voltage delivered by amplifier
V_{out} = Output voltage delivered to speaker
R_1 = Output Resistance of Amplifier
R_2 = Input Resistance of Speaker

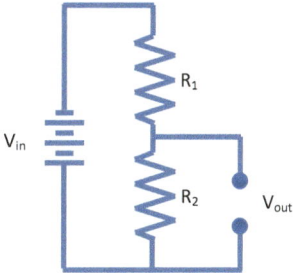

This expression is simple enough that it can be solved analytically. P_2 is either a maximum or minimum when $dP_2/dR_2 = 0$. The first step is to find the derivative of the power expression.

$$\frac{dP_2}{dR_2} = \frac{d}{dR_2}\left[\frac{R_2}{(R_1 + R_2)^2}V_{in}^2\right] = \left[\frac{1}{(R_1 + R_2)^2} - \frac{2R_2}{(R_1 + R_2)^3}\right]V_{in}^2 = 0 \qquad (2.16)$$

Thus

$$\frac{1}{(R_1 + R_2)^2} = \frac{2R_2}{(R_1 + R_2)^3} \qquad (2.17)$$

Multiplying through by $(R_1 + R_2)^2$ gives

$$R_1 + R_2 = 2R_2 \tag{2.18}$$

So $R_1 = R_2$, and power delivery is maximized when the output resistance of the supply is equal to the input resistance of the load.

To be sure that this is a maximum and not a minimum, we need to check that the curvature is negative. Remember that the curvature of a function is just its second derivative.

$$C = \frac{d^2}{dR_2^2}\left[\frac{R_2}{(R_1 + R_2)}V_{in}^2\right] = \left[\frac{6R_2}{(R_1 + R_2)^4} - \frac{4}{(R_1 + R_2)^3}\right]V_{in}^2 \tag{2.19}$$

Since $R_1 = R_2$

$$C = -\frac{V_{in}^2}{8R_2^3} \tag{2.20}$$

Since R_2 is always positive, C is always negative, and we are sure we have identified a maximum rather than a minimum.

2.4 Shortest Time Path for a Toy Car

Sometimes optimization problems take us a little bit by surprise. Let's consider the case of a toy car rolling down a segmented ramp under the force of gravity as shown in Fig. 2.9. Let's assume that the starting point, H_1, is 1 m above the end of the track and that length, L, is also 1 m. Finally, let's also assume that rolling resistance is negligible. Is there a value of H_2 that gets the car to the bottom of the track in a minimum time?

If you think about this problem in terms of energy, it's clear that all the potential energy the car has at the top of the track will be turned into kinetic energy at the bottom of the track [6]. So, as long as $H_2 < H_1$, the car will roll to the right, and it will

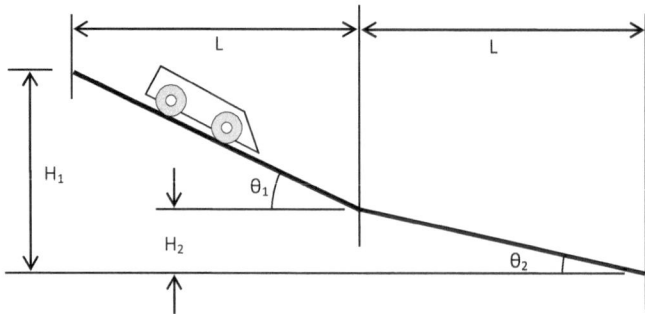

Fig. 2.9 Toy car rolling down a track with two segments

have the same speed at the end of the track, no matter what value H_2 has. The temptation is to then conclude that the time to finish is also the same no matter what the value of H_2.

To see that this isn't correct, imagine H_2 is just barely smaller than H_1. Imagine the difference is so small that the car accelerates very slowly and takes a long time to traverse the first section of the track. Clearly, this would take much longer than if the track was, say, straight ($H_2 = H_1/2$). Even though it may not be intuitive at first, it's clear that, while the car has the same speed at the finish, no matter what the value of H_2 might be, it is possible that different shapes of the track will give different transit times.

The objective function for this problem is the time to reach the end of the track, and the one design variable is H_2. The problem is then just to write out expressions for how long it takes to cover each section of track and add the two values.

We can start by defining the lengths of the two segments of track, L_1 and L_2.

$$L_1 = \sqrt{L^2 + (H_1 - H_2)^2} \quad \text{and} \quad L_2 = \sqrt{L^2 + H_2^2} \tag{2.21}$$

The acceleration of the car due to gravity is $g\sin\theta$ where

$$\sin\theta_1 = \frac{H_1 - H_2}{\sqrt{L^2 + (H_1 - H_2)^2}} \quad \text{and} \quad \sin\theta_2 = \frac{H_2}{\sqrt{L^2 + H_2^2}} \tag{2.22}$$

Assume the initial velocity is zero, so the time required to travel the first segment of the track is given by

$$L_1 = \frac{1}{2}a_1 t_1^2 \quad \text{or} \quad t_1 = \sqrt{\frac{2L_1}{a_1}} = \sqrt{\frac{2L_1}{g\sin\theta_1}} \tag{2.23}$$

Let's let V_1 be the velocity at the end of section 1. There are two convenient expressions for V_1.

$$V_1 = \sqrt{2a_1 L_1} = a_1 t_1 \tag{2.24}$$

Finally, the relationship between distance and time in the second section of the track is

$$t_2 = \frac{V_2 - V_1}{a_2} \quad \text{where} \quad V_2 = \sqrt{2gH_1} \tag{2.25}$$

The design variable for this problem is H_1, which is buried inside a_1 and a_2. Rather than unpack it all into a single analytical expression, the calculations and resulting design space are shown in Fig. 2.10. In this example, $L = H_1 = 1$ m.

The result here may be a bit surprising since our mathematical model predicts that the joint between the two segments should be below the end of the track. Thus, the

Toy Car on Segmented Ramp

$g := 9.81$ $L := 1$ $H_1 := 1$

Calculate Values for Segment 1   Calculate Values for Segment 2

$$L_1(H_2) := \sqrt{L^2 + (H_1 - H_2)^2}$$ $$L_2(H_2) := \sqrt{L^2 + H_2^2}$$

$$\sin\theta_1(H_2) := \frac{H_1 - H_2}{L_1(H_2)}$$ $$\sin\theta_2(H_2) := \frac{H_2}{L_2(H_2)}$$

$$a_1(H_2) := g \cdot \sin\theta_1(H_2)$$ $$a_2(H_2) := g \cdot \sin\theta_2(H_2)$$

$$t_1(H_2) := \sqrt{\frac{2 \cdot L_1(H_2)}{a_1(H_2)}}$$ $$V_1(H_2) := \sqrt{2 \cdot a_1(H_2) \cdot L_1(H_2)} \qquad V_2 := \sqrt{2 \cdot g \cdot H_1}$$

$$t_2(H_2) := \frac{V_2 - V_1(H_2)}{a_2(H_2)}$$

$$T(H_2) := t_1(H_2) + t_2(H_2)$$ <= Total time is sum of the times from the two segments

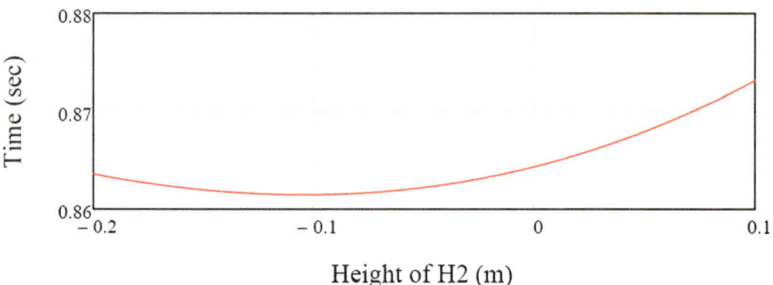

$H_2 := 0$ $\text{Minimize}(T, H_2) = -0.104$ <= Value of H_2 that gives minimum time

$T(-0.104) = 0.86148$ <= Minimum time in seconds

Fig. 2.10 Calculations for car on segmented ramp

car is actually decelerating along the second segment of the track. One valuable quality of optimization is that it can find counterintuitive solutions like this one.

When faced with a counterintuitive solution like this one, it is generally a good idea to do an approximate calculation to ensure that result is at least plausible. Imagine that $H_2 = 0$ so that the first portion of the track was at 45° and the second portion is horizontal. Working this out algebraically gives $T = 0.864$ s. This is close to the minimum value found for this problem and matches the point on the plot in

Fig. 2.10 for $H_1 = 0$. As a second check, assume $H_2 = H_1$ so that the track is straight. In this case, the acceleration is constant and is $a = g\sin\theta$, where $\theta = \tan^{-1}(1/2) = 26.565°$. Total time predicted by the math model in Fig. 2.10 is 1.00964 s and agrees with the results from the analytical expression for constant acceleration, $(L_1+L_2) = \frac{1}{2}at^2$.

2.5 Weight on a Snap-Through Spring

One of the most fascinating things about optimization is that it seems to be woven into the fabric of the universe. There are minimum principles underlying many of the physical laws we routinely use. One of those is that a structure will naturally take the shape that gives it the least possible total energy [7]. Here's an example showing how to use this minimum principle to solve a problem.

Let's say we have a vertical rod with a heavy collar around it. The collar has very good bearings, so it can move without friction. It is also supported by a spring as shown in Fig. 2.11.

Let's also say that the collar is now allowed to slide vertically under the force of gravity so it can compress the spring. Structures deform in ways that minimize energy, so the collar will move downward until it reaches a point of minimum total energy. The potential energy of the collar is given by $PE = -mgx$. Note that PE is negative since the positive direction is defined as down and the collar loses potential energy as it moves in that direction. The spring energy is given by $SE = \frac{1}{2}K\Delta L^2$, where ΔL is the change in length of the spring.

Finding the rest position of the collar means finding a value of x, the design variable that minimizes total energy, the objective function. This is equivalent to finding points at which the vertical component of the spring force equals the weight of the collar. In order to do this, we will need an expression that relates ΔL to x. The initial spring length is

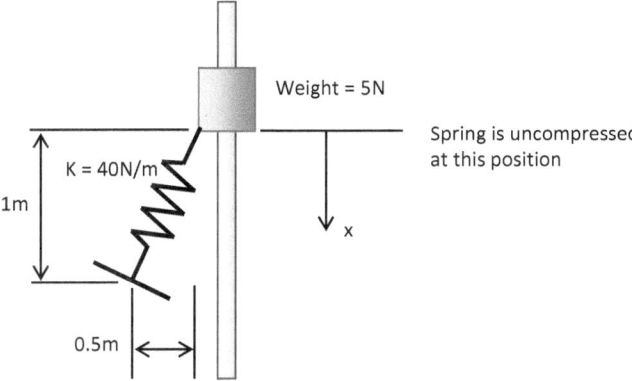

Fig. 2.11 Spring and collar assembly with spring uncompressed

$$L_0 = \sqrt{0.50^2 + 1.00^2} = 1.1180 \text{ m} \tag{2.26}$$

Spring length as a function of x is

$$L = \sqrt{0.50^2 + (1.00 - x)^2} \tag{2.27}$$

so the change in length is

$$\Delta L = L_0 - L = 1.118 \text{ m} - \sqrt{0.25 + (1.00 - x)^2} \tag{2.28}$$

Total energy as a function of x is

$$E = -mgx + \frac{1}{2}K\Delta L^2 = -5x + 20\left(1.1180 - \sqrt{0.25 - (1.00 - x)^2}\right)^2 \tag{2.29}$$

It is easy to plot the objective function for such a simple expression as shown in Fig. 2.12. The most important thing to note is that there are two minima, one at $x = 0.166$ m and another at $x = 2.150$ m. The first is called a local minimum. The slope of the objective function is zero, and its curvature is positive, so it meets the criteria to be called a minimum, but it is not the lowest point on the curve to satisfy those two requirements. The second point also has zero slope and positive curvature. Since the function value is the lower of the two, this point is called a global minimum.

Now we only have the problem of deciding what all this means physically. We will assume the weight is only allowed to move slowly, so kinetic energy is small enough to be ignored. If the weight is allowed to fall slowly, it will stop moving

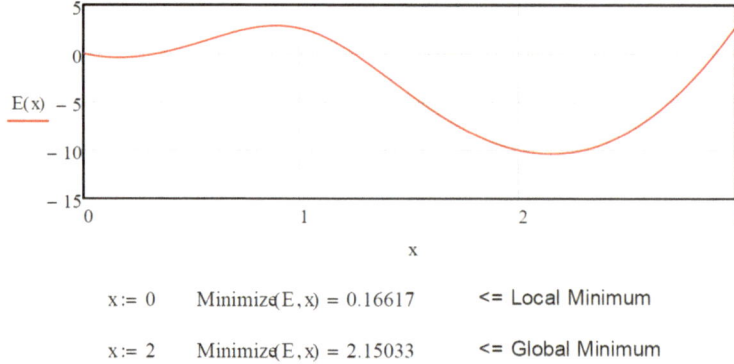

Weight on a Snap Through Spring

$$E(x) := -5x + 20\left[1.1180 - \sqrt{0.25 + (1.0 - x)^2}\right]^2$$

x := 0   Minimize(E, x) = 0.16617   <= Local Minimum

x := 2   Minimize(E, x) = 2.15033   <= Global Minimum

Fig. 2.12 Objective function of snap-through spring

Fig. 2.13 Global and local minima

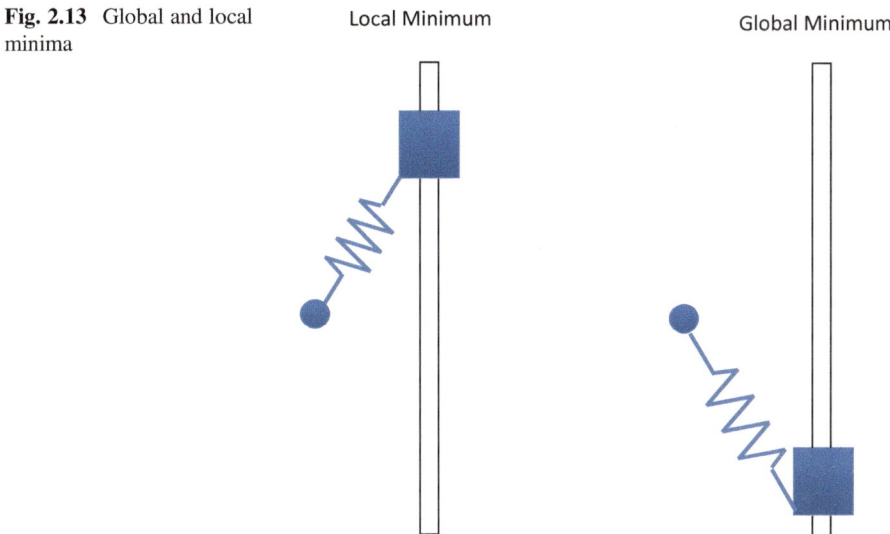

Local Minimum   Global Minimum

when $x = 0.166$ m. This is because the strain energy in the spring increases faster than the potential energy in the collar decreases. If we want the weight to settle at $x = 2.150$ m, there will have to be an additional external force to further compress the spring until it starts to stretch as shown in Fig. 2.13.

2.6   Solution Methods

Until now, all the numerical solutions have used the "minimize" and "maximize" commands in Mathcad. Clearly, the program is using some kind of algorithm to find maxima and minima. It's time to start learning about how optimization algorithms work. For now, we will limit the discussion to problems of a single variable. Later we'll advance to methods that work for more than one variable.

The single variable methods we'll cover are:

- Analytical solution
- Monte Carlo method
- Binary search method
- Quadratic approximation method
- Hybrid methods

A good way to study the different methods is to apply them all to the same few problems. In addition to the lifeguard problem, we'll use a short test function that has some useful features. The test function is shown in Fig. 2.14 along with some maxima and minima.

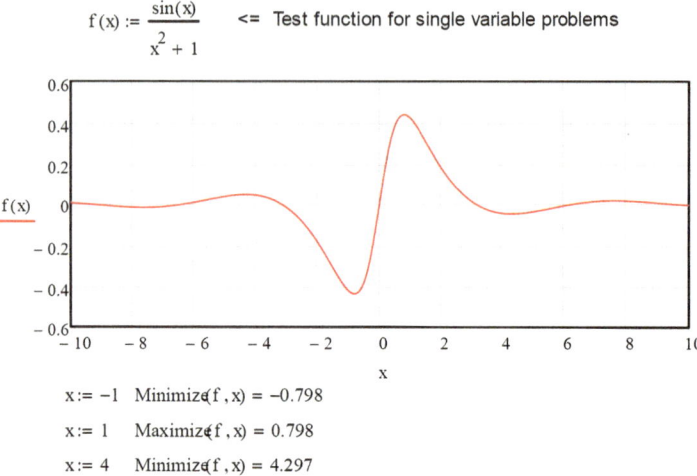

$$f(x) := \frac{\sin(x)}{x^2 + 1} \qquad <=\ \text{Test function for single variable problems}$$

$x := -1 \quad$ Minimize$(f, x) = -0.798$

$x := 1 \quad$ Maximize$(f, x) = 0.798$

$x := 4 \quad$ Minimize$(f, x) = 4.297$

Fig. 2.14 Test function for single variable methods

It is important to note that the goal of optimization algorithms is to find the minimum value of the objective function while calculating actual values of the objective function as few times as possible. Real optimization problems are often very computationally intensive, and efficiency is important. For this test function, it is easy to just plot it and look for the minima. In optimization terms, this is called mapping design space. The whole point of optimization is to find a minimum while not mapping any more of design space than is absolutely necessary.

In many cases, mapping design space would take far more time than is available. Imagine a case in which it takes 5 min to calculate the objective function for a single point in design space. Let's also imagine that it takes 1000 points to adequately map design space. That's 5000 min or about 83 h of computer time to do a single problem. It's easy to imagine that an efficient algorithm that would solve the same problem in a few dozen objective function evaluations would be a huge improvement. The bottom line is that we are trying to find a minimum value of the objective function while evaluating that objective function as few times as possible.

2.6.1   Analytical Solution

The analytical solution method is just what it sounds like. When there is a closed form expression for the objective function, it is possible to simply find the points at which the slope of the objective function is zero and the curvature (the second derivative) is positive. Finding points where the slope is zero is equivalent to finding roots of the derivative of the objective function as shown in Fig. 2.15. Note that the

$$f(x) := \frac{\sin(x)}{x^2 + 1} \qquad \text{<= Test function for single variable problems}$$

$$dfdx(x) := \frac{d}{dx}f(x) \rightarrow \frac{\cos(x)}{x^2 + 1} - \frac{2 \cdot x \cdot \sin(x)}{\left(x^2 + 1\right)^2} \qquad \text{<= Derivative of test function}$$

$$d2fdx(x) := \frac{d^2}{dx^2}f(x) \rightarrow \frac{8 \cdot x^2 \cdot \sin(x)}{\left(x^2 + 1\right)^3} - \frac{2 \cdot \sin(x)}{\left(x^2 + 1\right)^2} - \frac{\sin(x)}{x^2 + 1} - \frac{4 \cdot x \cdot \cos(x)}{\left(x^2 + 1\right)^2} \qquad \text{<= Curvature of test function}$$

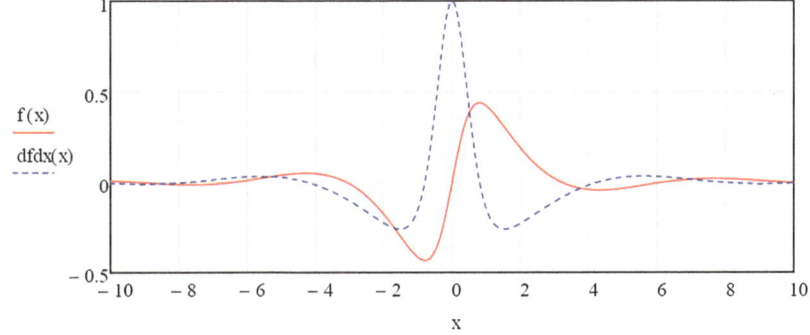

$x := -7$	$x := \text{root}(dfdx(x), x) = -7.601$	$d2fdx(x) = 0.017$
$x := -5$	$x := \text{root}(dfdx(x), x) = -4.297$	$d2fdx(x) = -0.052$
$x := -1$	$x := \text{root}(dfdx(x), x) = -0.798$	$d2fdx(x) = 0.972$
$x := 1$	$x := \text{root}(dfdx(x), x) = 0.798$	$d2fdx(x) = -0.972$
$x := 5$	$x := \text{root}(dfdx(x), x) = 4.297$	$d2fdx(x) = 0.052$
$x := 7$	$x := \text{root}(dfdx(x), x) = 7.601$	$d2fdx(x) = -0.017$

Fig. 2.15 Finding minima for test function analytically

roots in this example are found numerically. It is not usually possible to find minima by finding the roots of df/dx analytically so that the result is an algebraic expression.

As clean and unambiguous as this method is, it is generally of little use since few meaningful optimization problems have objective functions this simple. It is much more common for objective functions to be the result of extensive calculations. In these cases, we have to find a way to move numerically through design space as intelligently as we can.

2.6.2   Monte Carlo Method

The simplest approach to finding an approximate minimum value of the objective function without mapping all of the design space is to randomly pick a small number of points in design space, calculate the objective function at those points, and just pick the smallest one; this is a simple instance of the Monte Carlo method [8, 9]. Monte Carlo methods are used in a broad range of problems, and some implementations are quite sophisticated. They typically have a small number of features in common:

1. Define a domain of possible independent variables.
2. Generate a list of random points in the domain.
3. Calculate dependent variables for each random point.
4. Evaluate the results.

In the simplest form, these translate to:

1. Define independent variables that form design space.
2. Generate at list of random points in design space using whatever distribution you like.
3. Calculate objective function values for each point.
4. Select lowest objective function value.

This is, metaphorically, hitting the problem with big hammer. However, chances may be good that at least one of the randomly calculated points will yield a lower value of the objective function than the initial design. While it is a basic method, it is easy to implement, and it serves as a reminder that the perfect can be the enemy of the good. Figure 2.16 shows the Monte Carlo method applied to the test function.

There are 10 points randomly spaced along the *x* axis, and one is reasonably close to a local minimum. Of course, adding more points increases the chance of finding a point close to the global minimum. The problem is that selecting too large a number of points is essentially equivalent to mapping design space, and this is precisely what we wish to avoid.

This implementation of the Monte Carlo method is very simple, but it is also very robust; if you calculate enough points, you are likely to find a design more optimum that the one you started with. However, it's clear that the computational requirements are excessive for all but the simplest problems. A better approach is to employ a search algorithm that moves through design space in a more directed way. The most basic of these algorithms is called a binary search [10].

2.6.3   Binary Search

This is the simplest example of a search algorithm. It consists of two separate operations, first stepping forward to approximately locate the minimum and then

$f(x) := \dfrac{\sin(x)}{x^2 + 1}$ <= Test function for single variable problems   ORIGIN≡ 1

$N := 10$ $i := 1..N$ $xr_i := rnd(20) - 10$ <= Generate a list of N random
numbers between -10 and 10

$fr_i := f(xr_i)$ <= Generate list of corresponding
function values

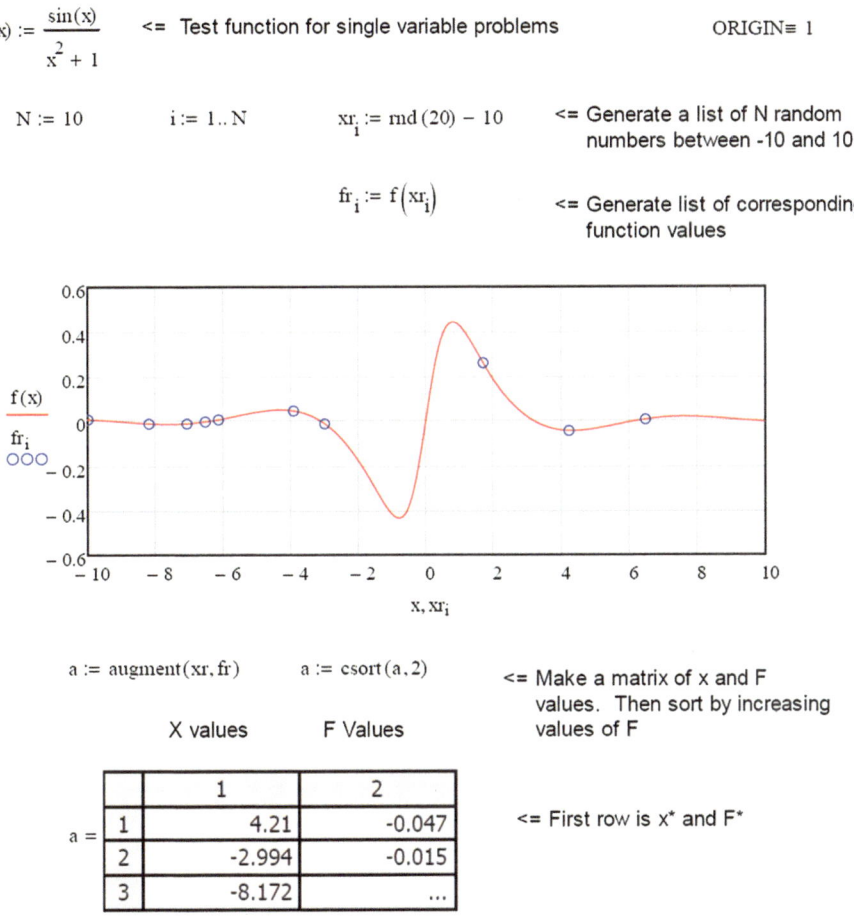

$a := augment(xr, fr)$ $a := csort(a, 2)$ <= Make a matrix of x and F
values. Then sort by increasing
values of F

		X values	F Values
		1	2
a =	1	4.21	-0.047
	2	-2.994	-0.015
	3	-8.172	...

<= First row is x* and F*

Fig. 2.16 Monte Carlo solution for test function

zooming in to locate it more precisely. The name comes from the fact that step sizes are successively cut in half to improve the estimate of the minimum value of the objective function. In more formal terms:

- Assign a step size, Δx, and move by this amount until the objective function stops decreasing.
- Cut the step size in half and evaluate the objective function at new points to see if there is a lower value. This step is repeated until the estimate of the minimum is sufficiently close to the exact value.

Let's apply the binary search method to the lifeguard problem where the objective function is

$$F(x) := \frac{1}{7} \cdot \sqrt{50^2 + x^2} + \frac{1}{2} \cdot \sqrt{50^2 + (100 - x)^2} \qquad \text{<= Define objective function}$$

$$\Delta x := 20 \qquad \text{<= Define initial step size}$$

$0 \cdot \Delta x = 0$	$F(0) = 63.045$	
$1 \cdot \Delta x = 20$	$F(\Delta x) = 54.863$	
$2 \cdot \Delta x = 40$	$F(2 \cdot \Delta x) = 48.199$	
$3 \cdot \Delta x = 60$	$F(3 \cdot \Delta x) = 43.173$	
$4 \cdot \Delta x = 80$	$F(4 \cdot \Delta x) = 40.403$	<= First estimate of minimum
$5 \cdot \Delta x = 100$	$F(5 \cdot \Delta x) = 40.972$	

$x_{star} := 4 \cdot \Delta x = 80$ <= x* designates the current estimate of x corresponding to the minimum value of the objective function

$F_{star} := F(x_{star}) = 40.403$ <= F* is the current estimate of the minimum objective function value

Fig. 2.17 First step in binary search example

$$T(x) = \frac{1}{7} \sqrt{50^2 + x^2} + \frac{1}{2} \sqrt{50^2 + (100 - x)^2} \qquad (2.30)$$

We already know that the objective function decreases as x increases from an initial value of $x = 0$, so we know we want to move to the right. We will make the initial step size large, say 20 m, and stop when the objective function begins to increase. Figure 2.17 shows this step in the calculation. It is standard practice to designate the estimate of the minimum as x^*.

We have identified an approximate minimum at $x = 80$ m using the first part of the binary search algorithm. We can guess that the actual minimum is somewhere near this point, but we need to reduce the step size and keep looking. There are two possibilities to account for as shown in Fig. 2.18. The dots are points that have already been calculated, and the point in the center is the current estimate of the minimum. These are the last three points calculated in Step 1 above. We assume that the objective function doesn't have several minima between the points we've already calculated.

One way to account for these two possibilities is to cut Δx in half and calculate one additional point on each side of the current estimate of the minimum as shown in Fig. 2.19. Then, a new minimum can be chosen.

One of two things will happen at this stage. Either one of the two new points will be a new minimum or the original point will remain a minimum. In the first case, one of the new points is the new minimum. This step is shown in Fig. 2.20.

With the new estimates of x^* and F^* in hand, we can cut the step size in half again and repeat the shown in Fig. 2.21. The estimate for x^* is now 85 m, very close to what we already know is the right answer.

Clearly, this process can go on forever unless we create some exit criteria. One exit criteria is to stop when Δx goes below a preset value. Another is to stop when F^* stops changing, though it might remain unchanged for several iterations if the step size is large. We will continue for one more iteration to see what happens, shown in Fig. 2.22.

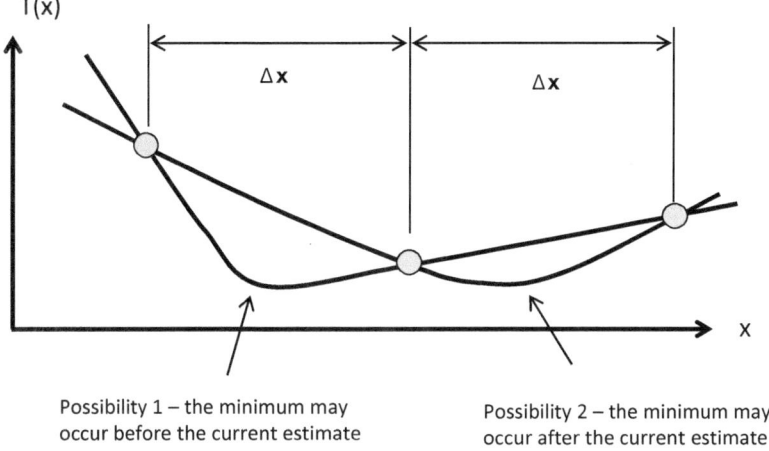

Fig. 2.18 Possible Locations of Minima

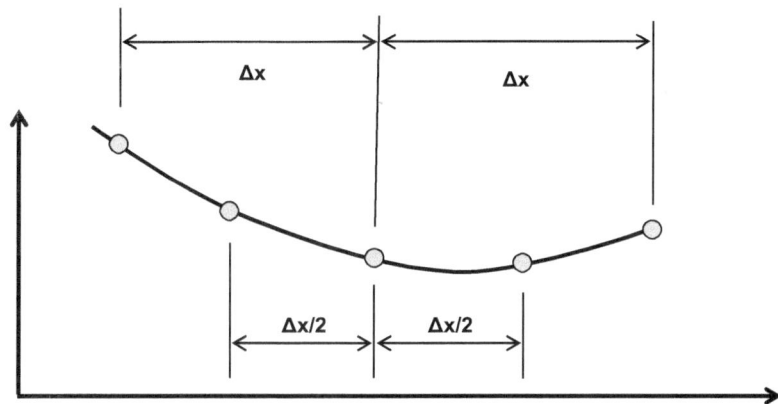

Fig. 2.19 Cutting the step size in binary search

The estimate of x^* is now very close to what we know is the right answer. In fact, F^* is correct to five significant figures, but, in practice, we wouldn't know that. We would only know that we were within 2.5 m of the right answer. If further precision was needed, we need only continue the algorithm until a suitable exit condition is reached. Finally, Fig. 2.23 shows a pseudo flow chart for the binary search algorithm using Δx_{min} as the exit condition.

Binary search is very robust in that it doesn't get lost and will always find a minimum. However, it can require a large number of objective function evaluations. Remember the goal is to do as few evaluations as possible. A method that can, under the right circumstances, find a minimum more quickly is the quadratic approximation method.

$$\Delta x := \frac{\Delta x}{2} = 10 \qquad \text{<= Cut } \Delta x \text{ in half}$$

$$x_{star} - \Delta x = 70 \qquad F\!\left(x_{star} - \Delta x\right) = 41.444$$

$$x_{star} = 80 \qquad F\!\left(x_{star}\right) = 40.403$$

$$x_{star} + \Delta x = 90 \qquad F\!\left(x_{star} + \Delta x\right) = 40.203 \qquad \text{<= New estimate of minimum}$$

$$x_{star} := x_{star} + \Delta x = 90$$

$$F_{star} := F\!\left(x_{star}\right) = 40.203$$

Fig. 2.20 Effect of cutting step size in half

$$\Delta x := \frac{\Delta x}{2} = 5 \qquad \text{<= Cut } \Delta x \text{ in half}$$

$$x_{star} - \Delta x = 85 \qquad F\!\left(x_{star} - \Delta x\right) = 40.189 \qquad \text{<= New estimate of minimum}$$

$$x_{star} = 90 \qquad F\!\left(x_{star}\right) = 40.203$$

$$x_{star} + \Delta x = 95 \qquad F\!\left(x_{star} + \Delta x\right) = 40.461$$

$$x_{star} := x_{star} - \Delta x = 85$$

$$F_{star} := F\!\left(x_{star}\right) = 40.189$$

Fig. 2.21 Reducing step size by half again

2.6.4   Quadratic Approximation Method

It is generally difficult to find the minimum of an objective function when we don't know the exact expression for that function. What if we could approximate the objective function with a parabola? Finding the minimum value of a parabola is very easy, so the utility of this idea depends on how well the parabola approximates the actual function [11].

Ideally, we could start with a quadratic approximation near some starting point and then find an approximate minimum. The next step would be to do another quadratic approximation, this time about the new estimate of the minimum, and then update the estimate of the minimum by finding the minimum of the new parabola. This algorithm can be repeated as many times as needed for the estimated minimum to converge to the minimum of the objective function.

$$\Delta x := \frac{\Delta x}{2} = 2.5 \quad \text{<= Cut } \Delta x \text{ in half}$$

$x_{star} - \Delta x = 82.5 \qquad F\left(x_{star} - \Delta x\right) = 40.268$

$x_{star} = 85 \qquad\qquad F\left(x_{star}\right) = 40.189$

$x_{star} + \Delta x = 87.5 \qquad F\left(x_{star} + \Delta x\right) = 40.166 \qquad \text{<= New estimate of minimum}$

$$x_{star} := x_{star} + \Delta x = 87.5$$

$$F_{star} := F\left(x_{star}\right) = 40.166$$

Fig. 2.22 Reducing step size a third time

Let's start with the lifeguard problem using calculations we've already made. We will use the first three values of x and the corresponding values of F to define a parabola. In practice, this will be a system of three linear equations and three unknowns.

$$\begin{aligned} F_1 &= ax_1^2 + bx_1 + c \\ F_2 &= ax_2^2 + bx_2 + c \\ F_2 &= ax_3^2 + bx_3 + c \end{aligned} \qquad (2.31)$$

Note that the unknowns in this set of equations are a, b, and c. Defining the parabola that passes through the three known points means finding these unknowns. Finding the minimum value of the parabola, $P(x)$, is just a matter of setting $dP/dx = 0$ and solving for x. Doing so results in $x^* = -b/2a$. This calculation is shown in Fig. 2.24.

The very first iteration requires only three objective function evaluations and gets within about 30 m of the correct answer. In practice, we won't know where the answer is beforehand, so the next step is to pick a point on either side of our current estimate of x^* and do another parabolic fit. Since the starting point is closer to the actual minimum, one hopes it will more precisely estimate the exact minimum. Figure 2.25 shows that the second estimate of x^* is 83.345 m. The estimate is now close to the true minimum, and there have only been six objective function evaluations.

Since the change in x^* was larger than Δx, it makes sense to leave Δx unchanged. Doing this, the third estimate of x^* is 86.392. This is less than 1 m from the exact minimum.

The change in x^* is now much smaller than Δx and it makes sense to reduce Δx. There are several possibilities. One is to cut Δx in half and another is to make it equal to the change in x^* from the previous iteration. Remember that there is no reason to think the objective function is actually parabolic. Rather, we are assuming it is

Select Δx_{min} as an exit condition
Select ΔF_{min} as an exit condition
Select x_0
Select Δx

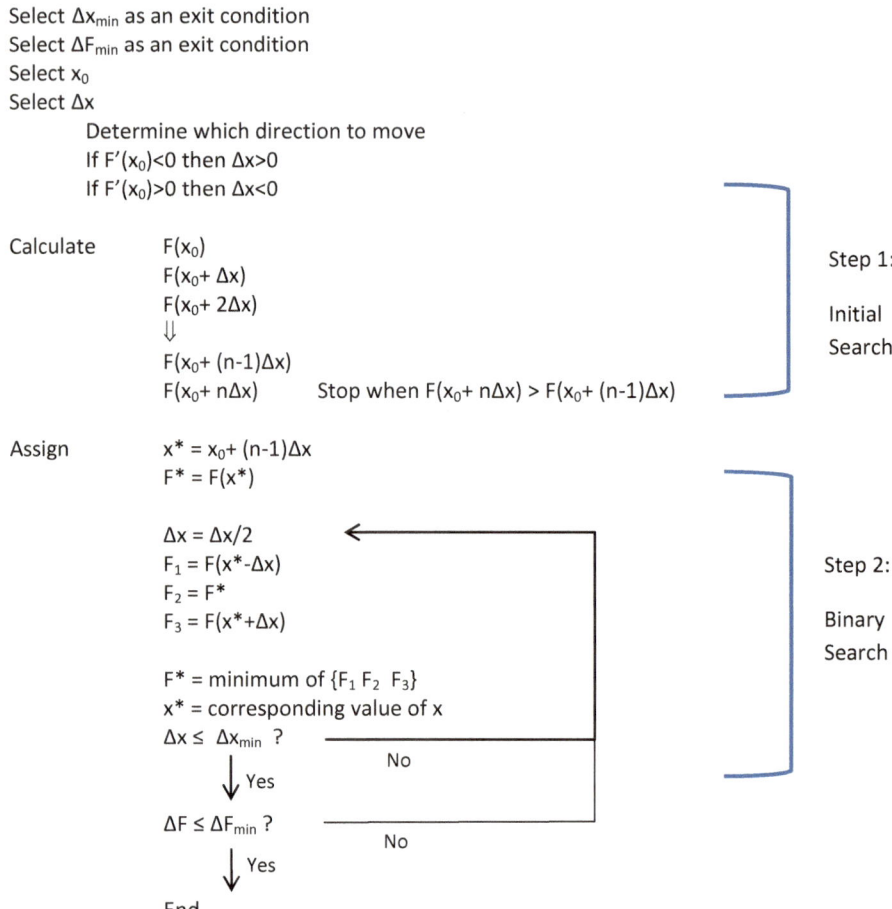

Fig. 2.23 Pseudo flow chart for binary search

approximately parabolic over a narrow range. Cutting Δx in half ($\Delta x = 10$ m) and doing a fourth iteration resulted in $x* = 86.99$ m, about 0.2 m from the exact value of $x*$.

There are two obvious concerns with the parabolic approximation. The first is that the user must pick an appropriate value of Δx. Physical conditions and prior knowledge of the problem made it easy in this case. In practice, the user will need to use what information about the problem might be available. Poor choices of Δx may cause slower convergence.

The other, more serious, concern is that the parabolic approximation only has a minimum value if the curvature is positive. Curvature is the second derivative, so positive curvature requires that the coefficient 'a' must be positive. As an example, go back to the test function used earlier in this chapter and use a starting point of

$$F(x) := \frac{1}{7} \cdot \sqrt{50^2 + x^2} + \frac{1}{2} \cdot \sqrt{50^2 + (100 - x)^2}$$

Start with three points

$$x_1 := 0 \qquad F_1 := F(x_1) = 63.04456$$
$$x_2 := 20 \qquad F_2 := F(x_2) = 54.863$$
$$x_3 := 40 \qquad F_3 := F(x_3) = 48.19857$$

Form a system of three equations

$$m := \begin{pmatrix} x_1^2 & x_1 & 1 \\ x_2^2 & x_2 & 1 \\ x_3^2 & x_3 & 1 \end{pmatrix} = \begin{pmatrix} 0 & 0 & 1 \\ 400 & 20 & 1 \\ 1600 & 40 & 1 \end{pmatrix} \qquad v := \begin{pmatrix} F_1 \\ F_2 \\ F_3 \end{pmatrix} = \begin{pmatrix} 63.04456 \\ 54.863 \\ 48.19857 \end{pmatrix}$$

Solve for a, b and c

$$\begin{pmatrix} a \\ b \\ c \end{pmatrix} := m^{-1} \cdot v = \begin{pmatrix} 0.0019 \\ -0.44701 \\ 63.04456 \end{pmatrix}$$

Define the parabola $\qquad P(x) := a \cdot x^2 + b \cdot x + c$

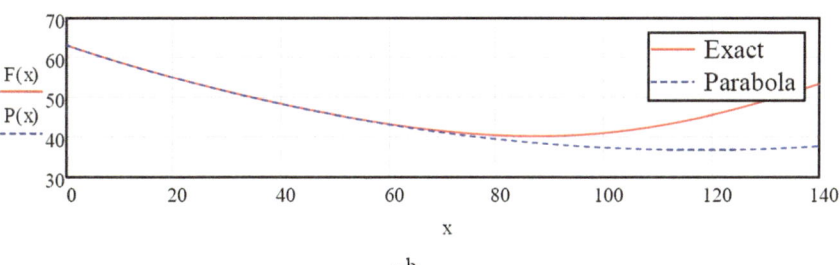

Minimum occurs at x = -b/2a $\qquad x_{star} := \dfrac{-b}{2a} = 117.85582 \qquad$ <= Estimate of minimum, x*

Fig. 2.24 Parabolic approximation for lifeguard problem

$x = -4$ m and $\Delta x = 1$ m. The resulting parabola has negative curvature, so the parabolic approximation method fails in that case.

Finally, Fig. 2.26 shows a pseudo flow chart for the parabolic approximation method. This implementation cuts the step size in half if the change in x* from the previous iteration is less than the current value of Δx.

2.6.5 Hybrid Methods

We have seen so far that the binary search method is robust in that it always finds the minimum, but it may not be very efficient. Conversely, the parabolic approximation

Second Iteration of Parabolic Approximation

$\Delta x := 20$

$x_1 := x_{star} - \Delta x = 97.85582$ $F_1 := F(x_1) = 40.72152$

$x_2 := x_{star} = 117.85582$ $F_2 := F(x_2) = 44.83539$

$x_3 := x_{star} + \Delta x = 137.85582$ $F_3 := F(x_3) = 52.3061$

Form a system of three equations

$$m := \begin{pmatrix} x_1^2 & x_1 & 1 \\ x_2^2 & x_2 & 1 \\ x_3^2 & x_3 & 1 \end{pmatrix} = \begin{pmatrix} 9575.76 & 97.86 & 1 \\ 13889.99 & 117.86 & 1 \\ 19004.23 & 137.86 & 1 \end{pmatrix} \qquad v := \begin{pmatrix} F_1 \\ F_2 \\ F_3 \end{pmatrix} = \begin{pmatrix} 40.72152 \\ 44.83539 \\ 52.3061 \end{pmatrix}$$

Solve for a, b and c

$$\begin{pmatrix} a \\ b \\ c \end{pmatrix} := m^{-1} \cdot v = \begin{pmatrix} 0.0042 \\ -0.69944 \\ 68.98569 \end{pmatrix} \qquad P(x) := a \cdot x^2 + b \cdot x + c$$

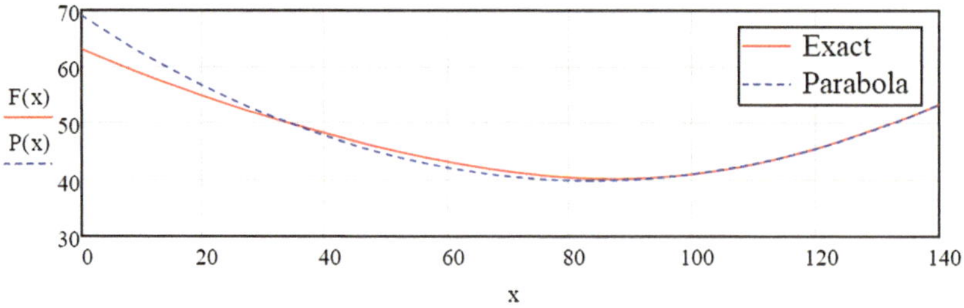

Minimum occurs at x = -b/2a $x_{star} := \dfrac{-b}{2a} = 83.34541$ <= This is the second estimate of x*

Fig. 2.25 Second iteration of parabolic approximation

method is more efficient when it works but fails when the curvature is negative. An obvious response would be to combine methods into a hybrid approach. Consider what would happen if the initial search portion of the binary search algorithm was to be combined with a parabolic approximation. The calculation is shown in Fig. 2.27. The first estimate of x^* from the binary search gives $x^* = 80$ m. The second estimate from the parabolic approximation gives $x^* = 86.592$ m.

In this case, a fair estimate of the minimum (about 0.6 m from the exact value) was possible with only six objective function evaluations. The curvature of $T(x)$ is positive in the region of interest, so the three points used to define the parabola will also give a positive curvature.

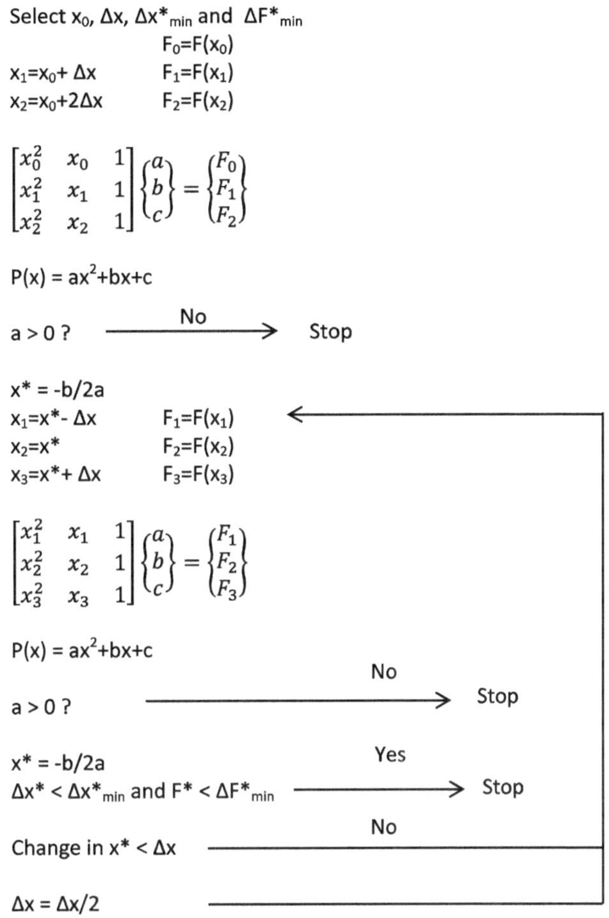

Fig. 2.26 Pseudo flow
chart for parabolic
approximation method

Select x_0, Δx, Δx^*_{min} and ΔF^*_{min}

$$F_0 = F(x_0)$$

$x_1 = x_0 + \Delta x$ $F_1 = F(x_1)$

$x_2 = x_0 + 2\Delta x$ $F_2 = F(x_2)$

$$\begin{bmatrix} x_0^2 & x_0 & 1 \\ x_1^2 & x_1 & 1 \\ x_2^2 & x_2 & 1 \end{bmatrix} \begin{Bmatrix} a \\ b \\ c \end{Bmatrix} = \begin{Bmatrix} F_0 \\ F_1 \\ F_2 \end{Bmatrix}$$

$P(x) = ax^2 + bx + c$

$a > 0$? $\xrightarrow{\text{No}}$ Stop

$x^* = -b/2a$

$x_1 = x^* - \Delta x$ $F_1 = F(x_1)$ ⟵

$x_2 = x^*$ $F_2 = F(x_2)$

$x_3 = x^* + \Delta x$ $F_3 = F(x_3)$

$$\begin{bmatrix} x_1^2 & x_1 & 1 \\ x_2^2 & x_2 & 1 \\ x_3^2 & x_3 & 1 \end{bmatrix} \begin{Bmatrix} a \\ b \\ c \end{Bmatrix} = \begin{Bmatrix} F_1 \\ F_2 \\ F_3 \end{Bmatrix}$$

$P(x) = ax^2 + bx + c$

$a > 0$? $\xrightarrow{\text{No}}$ Stop

$x^* = -b/2a$ Yes

$\Delta x^* < \Delta x^*_{min}$ and $F^* < \Delta F^*_{min}$ $\longrightarrow$ Stop

         No

Change in $x^* < \Delta x$ ⟶

$\Delta x = \Delta x/2$

2.7 Finding Roots Using Minimization

Finding roots of equations is an important task, and, as a result, there are many
different algorithms for finding the roots of equations. One simple approach is to use
the ideas from unconstrained optimization.

    Finding the root of a function, $f(x)$, means finding a value of x so that $f(x) = 0$. Say
that x is chosen so that $f(x)$ is not exactly zero. Then $f(x) = \varepsilon$ and finding a root of the
equation means finding a value of x so that ε is zero. There is not necessarily a
minimum value of ε since it can be negative. Instead, we can square both sides of the
equation so that $f(x)^2 = \varepsilon^2$. If the equation has at least one root, we know that the
minimum value of ε is zero and occurs when $f(x) = 0$. This is equivalent to
minimizing $f(x)^2$.

    As an example, let's find the roots of this equation

$$F(x) := \frac{1}{7} \cdot \sqrt{50^2 + x^2} + \frac{1}{2} \cdot \sqrt{50^2 + (100 - x)^2}$$

Start Binary Search with $\Delta x = 20m$

$x_1 := 0$	$F_1 := F(x_1) = 63.04456$	$x_4 := 60$	$F_4 := F(x_4) = 43.17312$
$x_2 := 20$	$F_2 := F(x_2) = 54.863$	$x_5 := 80$	$F_5 := F(x_5) = 40.40294$ <= Initial estimate of x*
$x_3 := 40$	$F_3 := F(x_3) = 48.19857$	$x_6 := 100$	$F_6 := F(x_6) = 40.97191$

Form a system of three equations using last three values of x

$$m := \begin{pmatrix} x_4^2 & x_4 & 1 \\ x_5^2 & x_5 & 1 \\ x_6^2 & x_6 & 1 \end{pmatrix} = \begin{pmatrix} 3600 & 60 & 1 \\ 6400 & 80 & 1 \\ 10000 & 100 & 1 \end{pmatrix} \qquad v := \begin{pmatrix} F_4 \\ F_5 \\ F_6 \end{pmatrix} = \begin{pmatrix} 43.17312 \\ 40.40294 \\ 40.97191 \end{pmatrix}$$

Solve for a, b and c

$$\begin{pmatrix} a \\ b \\ c \end{pmatrix} := m^{-1} \cdot v = \begin{pmatrix} 0.00417 \\ -0.72286 \\ 71.51859 \end{pmatrix} \qquad P(x) := a \cdot x^2 + b \cdot x + c$$

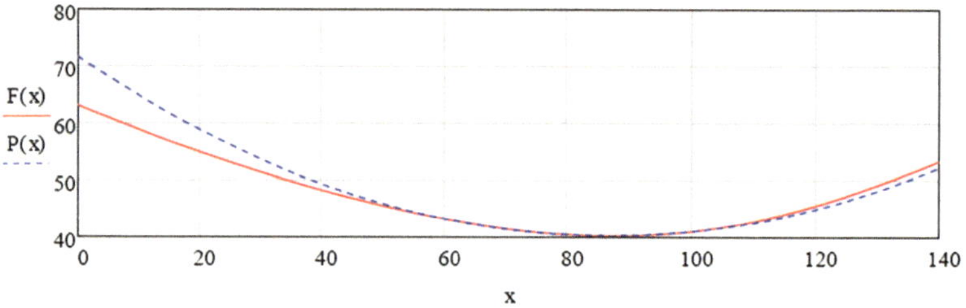

Minimum occurs at x = -b/2a $x_{star} := \dfrac{-b}{2a} = 86.59211$ <= This is the first estimate of x*

Fig. 2.27 Sample calculation for hybrid method

$$f(x) = \frac{(x - 2)^2}{2} - \sqrt{x^2 + 2} \tag{2.32}$$

Finding the roots of this equation is equivalent to finding the minimum values of the squared function

$$\text{minimize} \quad \left[\frac{(x - 2)^2}{2} - \sqrt{x^2 + 2} \right]^2 = \varepsilon^2 \tag{2.33}$$

Figure 2.28 shows that the roots of the test function correspond to the minima of the squared remainder, ε^2.

A related approach is possible when the roots of a function are complex. Consider the parabola $y = x^2 + 4$ where the roots are $x = 2i$ and $x = -2i$. The real and

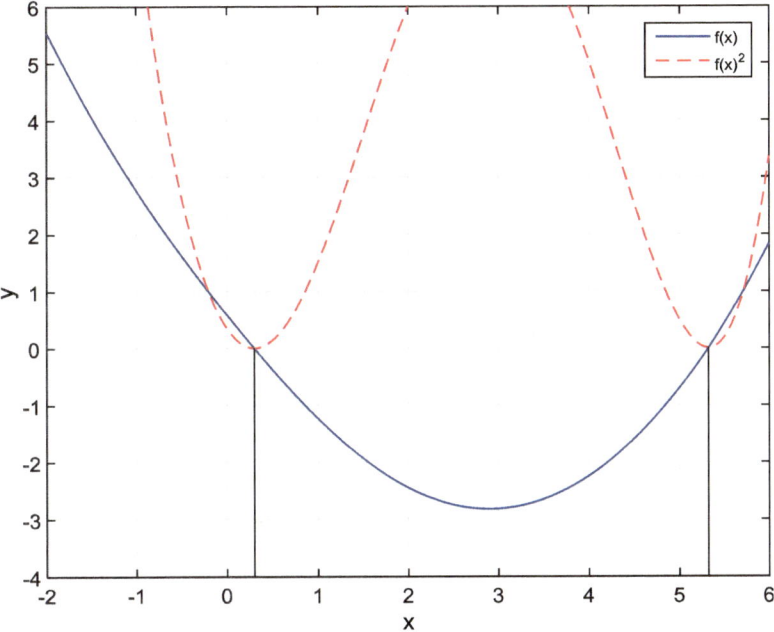

Fig. 2.28 Roots of test function, $x_1 = 0.29964$ and $x_2 = 5.3173$

imaginary parts of x can be treated as independent variables, and the absolute value of the function, $|f|$, becomes the dependent variable. Figure 2.29 shows the resulting surface plot.

2.8   Optimizing a Process: Ocean Shipping Routes

A valuable way of learning how optimization works is to consider practical applications that have worked well. The mathematical tools of optimization can be very powerful but are of little help unless the user is able to recognize an optimization problem and bring the mathematical tools to bear. Among the many choices of successful applications in optimization, one that stands out is the process of shipping goods across the oceans. The goal is usually to move goods from one place to another as cheaply as possible, so cost is the objective function.

People have been moving products and raw materials by sea for thousands of years, and transoceanic trade has been a huge boost to national economies [12]. One of the oldest shipwrecks ever found is off the coast of southern Greece, near the island of Dokos. The wreck has been dated to between 2700 and 2200 B.C. The site was littered with pottery, amphorae, and other manufactured goods, so this was likely a trading vessel [13]. It's thus clear that people have been using ships to move goods for more than 4000 years.

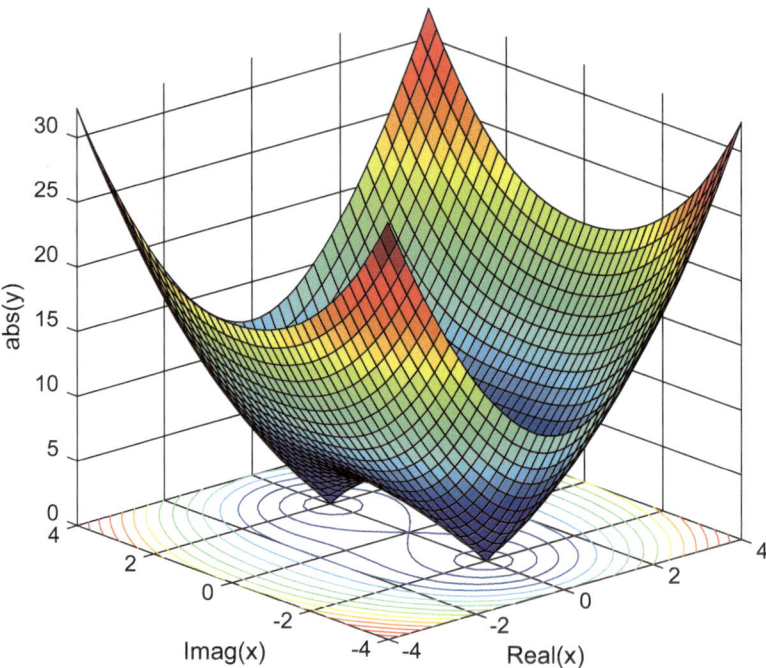

Fig. 2.29 Surface plot showing complex roots

The story of transporting goods across oceans is, in part, the story of the development of better ships. By the time that Europeans began to seriously explore the world, ships had evolved far from the ones used by the ancient Greeks. Figure 2.30 shows a replica of the galleon "Batavia," an armed merchant vessel commissioned by the Dutch East India Company and originally launched in 1628. Galleons were stout and seaworthy, capable of crossing oceans while carrying enough cargo to make them profitable.

Sailing ships continued to evolve, reaching their pinnacle right before the development of steam engines small enough and reliable enough to drive ships across oceans. Perhaps the most well-known of these latter sailing ships were the tea clippers, optimized to bring compact but valuable cargos of tea to England from China. Finally, though, steam supplanted sail. While a few commercial sailing ships were still in service as late as 1920, the general conversion to steam power started about the end of the American Civil War (1865).

It is surprising how little the process of moving goods across the oceans changed between ancient times and the mid-twentieth century. The ships themselves changed from sail to steam and from wood to steel, but the method of handling the cargo evolved very little. Basically, the cargo was packaged in small, often nonstandardized containers like boxes and barrels and then loaded into the holds by laborers called stevedores. The process was lengthy and labor-intensive, often

Fig. 2.30 A replica of the galleon "Batavia" in Lelystad, the Netherlands. (Wikimedia Commons, image is in the public domain)

taking days to load or unload a ship. Figure 2.31 shows an example of loading loose cargo.

The beginnings of truly systemic improvements in moving cargo across oceans came during WWII when transoceanic shipping literally kept nations alive. The problem was brutal but simple – America's allies in Europe desperately needed food, medicine, munitions, and other manufactured goods. If they couldn't get these things, they might be forced to surrender. The USA could produce the needed goods but had to ship them across the Atlantic Ocean in spite of a submarine blockade that sank many ships. Part of the solution was to produce cargo ships on a scale never before seen and to do this faster than they could be sunk by submarines. These new ships, called Liberty ships [14], were mass produced using modular techniques that are now familiar. Figure 2.32 shows a Liberty ship under way.

Liberty ships were simple, dependable vessels that were easy to produce quickly. They were welded together from prefabricated sections and could be launched a few weeks after the keel was laid. This was possible, in part, because of standardization and made possible by using just a few standard designs. In spite of frightful losses, more than 2400 of them survived the war.

Even though the ships themselves were standardized and produced in quantity, the cargo was still generally loaded as it had always been, largely by hand and in small, nonstandardized containers. Liberty ships were designed to hold about 10,000

Fig. 2.31 Laborers loading bales of wool at the Wharf in Brisbane, circa 1931. (Wikimedia Commons, image is in the public domain)

tons but were often loaded with more than that. It took days to load or unload one, time when it was effectively out of service. Figure 2.33 shows a net full of boxes being unloaded from a Liberty ship during WWII. Note the men in the lower right portion of the picture who are loading the next cargo net by hand.

The first part of the puzzle then was the widespread availability of standardized ships. The second part was standardized containers [15]. There had been local attempts to standardize shipping containers, starting well before WWII, but they never entered widespread use. In 1952, the US Army began describing their standardized containers using the term "CONEX," short for "container express." These were first used during the Korean War and, by the Vietnam War, the majority of materials and supplies were shipped by CONEX. The US Department of Defense defined an 8 foot by 8 foot cross-sectional container in multiples of 10 foot lengths for military use, and it was rapidly adopted for shipping.

The last step in the process of standardizing containers came in 1955 when Malcom McLean, a former trucking company owner, and Keith Tantlinger, an engineer, developed the modern intermodal freight container. McLean gave the patented design to the industry to help start the international standardization of shipping containers [16].

Fig. 2.32 A Liberty ship at sea. (Wikimedia Commons, image is in the public domain)

In 1956, McLean put 58 standard containers aboard a refitted WWII oil tanker called the Ideal X for his first containerized shipment. While it wasn't the first ship designed for containerized cargo, it was the predecessor of the thousands of container ships that now ply the seas. The containers were loaded once by the customer, closed, and then opened again only when they reached their destination, so there was no need to handle loose cargo. Since all containers are effectively the same (there are a few different lengths), a purpose-built crane can be used to handle them. The result was a huge decrease in the time and people required to load and unload ships and a proportional decrease in cost – the objective function.

So how successful has this development been? By 2009, approximately 90% of non-bulk cargo worldwide was moved by containers stacked on container ships. Sizes of container ships are now expressed in 20-foot equivalent units (TEUs).

As standardized containers greatly improved efficiency in shipping, the next question was how to make the ships themselves more efficient. Efficiency is defined as the cost required to transport a container, so increased efficiency means reduced shipping cost. International trade requires hundreds of millions of container trans-shipments each year, so even a small decrease in the cost to ship a container can really matter.

Fig. 2.33 A Liberty ship being unloaded in Algiers in 1943. (Imperial War Museum)

It turns out that large ships can move containers more cheaply than small ships, and the recent history of container ships has been one of a dramatic increase in their size. As I write this, the largest container ship afloat is the MSC Oscar (Fig. 2.34). It is 395.4 m (1297 ft) long and rated at 19,224 TEU. For comparison, the Ideal X was 160 m (524 ft) long and (assuming 58 containers, each 33 ft long) may have been rated at less than 200 TEU. By the time you read this, it is likely that the largest container ship will be larger than even the MSC Oscar. The current terminology for ships rated at 14,500 TEU is an Ultra Large Container Vessel (ULCV).

So that's it then, right? Standardized shipping containers minimize the cost of handling cargo, and ultra large ships minimize the cost of moving them. That's the whole story? Well, not yet. There is still the decision of how to route these ships as they move from port to port.

There are opportunities to optimize the shipping process by selecting routes that further reduce costs [17, 18]. Imagine the simplest case in which a ship takes on its entire load at one port and delivers it all to another without any intermediate stops. Then, the objective might be to find the shortest route with the assumption that this will minimize fuel costs.

If the route is relatively direct, say, from the east coast of China to the west coast of the USA, the decision might be a simple one, but that's often not the case. Assume that the route is from the east coast of China to the east coast of the USA. Now, the

Fig. 2.34 The MSC Oscar, an Ultra Large Container Vessel. (Wikimedia Commons, image is in the public domain)

problem is more difficult for the obvious reason that the North American continent is in the way.

One obvious choice is to go through the Panama Canal. However, the locks were designed more than 100 years ago, and the designers assumed much smaller ships than are now generally being built. The largest ships capable of traversing the canal are designated as Panamax. These are rated at 3000–5100 TEU and cannot be wider than 32.2 m (106 ft). Panamax ships are very much smaller than the largest container ships now being launched. However, it might be that the advantage of the short route through the Panama Canal offsets the increased cost of using a ship small enough to traverse the locks. Figure 2.35 shows just such a vessel moving through one of the locks on the Panama Canal.

So, the really large and efficient ships can't get through the canal. One choice would be to take a southern route around Cape Horn at the southern tip of South America. However, this is a notoriously dangerous route, and many ships have sunk there. It is also very far south, near the coast of Antarctica.

Another choice is to go west from China, through the Indian Ocean and the Suez Canal. This avoids the long and treacherous trip around Cape Horn but is only possible if the ship can fit through that canal. There is a size limitation that naval architects call Suezmax, analogous to Panamax. A Suezmax ship is currently the neighborhood of 10,000–12,000 TEU, so ULCV ships can't traverse that canal either.

If there is enough ship traffic, the overall cost of the system might be reduced by improvements to the infrastructure rather than just to the ships themselves. As this is being written, Panama is expanding the lock system in the canal to accommodate larger ships. The new locks will allow passage by ships 49 m (161 ft) wide. These

Fig. 2.35 A Panamax ship moving through the Miraflores Locks on the Panama Canal. (Wikipedia Commons, images is in the public domain)

ships, designated New Panamax, are up to 14,500 TEU, so most current ships will fit.

Of course, some less obvious choices might work. Shipping containers are intermodal, which means they will fit on trains and trucks as well as ships. If the goal is to get containerized cargo from the east cost of China to the east coast of the USA, it is at least conceivable that the lowest cost option is taking it to the west coast of the USA by ship by sea and then across the USA by train or truck.

Shipping companies worldwide have been trying to optimize the system of moving freight by sea for centuries, with some of the really significant changes happening since the beginning of the twentieth century – the switch from coal to oil, the opening of the Panama canal, and the adoption of containerized freight. There is no single picture or plot that can capture the complexity of the problem or the nuances of the resulting solutions. However, a map showing the sea routes around the world and how heavily they are used is helpful (Fig. 2.36).

This figure shows that one of the busiest shipping routes is down the coast of China, directly across the Indian Ocean and through the Suez Canal. It's also clear that the route the Cape of Good Hope at the southern tip of Africa is popular, but essentially nobody makes the trip around Cape Horn.

The point of this section is that people have been working to improve the process of moving goods by ship for thousands of years. In the last hundred years or so, there have been major advances that have reduced costs to the point that transoceanic shipping is a major component of many national economies. This is an optimization

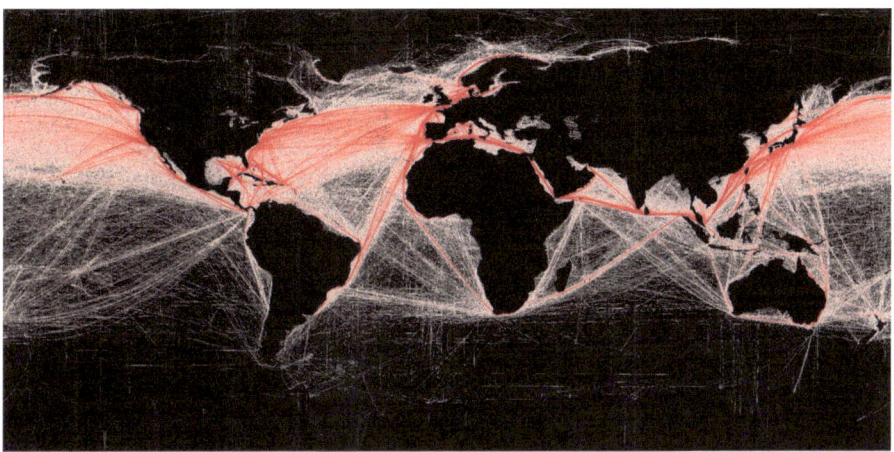

Fig. 2.36 Shipping routes around the world. (Wikimedia Commons, image is in the public domain)

problem on a global scale, and efforts to find more optimal solutions to the problem are certainly continuing.

What we have right now is not the best possible solution to the problem of shipping goods across the oceans of the world. It is only the best solution we've yet been able to find. It's a safe bet that more optimal solutions are coming.

2.9 MATLAB Examples

MATLAB is a very powerful and well-developed computational language, so there are many ways to solve most problems. The easiest way to solve the lifeguard problem in MATLAB is to use a function from the Optimization Toolbox. One of these is called "fminunc" and, as the name suggests, minimizes unconstrained functions. The code follows:

```matlab
function lifeguard1

% This function solves the one variable lifeguard problem
% using fminunc

% Define the objective function using an anonymous function

T = @(x)  1/7*sqrt(50^2+x^2)+1/2*sqrt(50^2+(100-x)^2);

x_min=fminunc(T,0); % Call fminunc using t=0 as a starting point
t_min=T(x_min);
disp(['The minimum time is ' num2str(t_min) ' seconds and occurs at x = ' num2str(x_min) ' meters'])
return
```

The output to the command window follows. Warnings are just that they are not errors. In this case, the warning is just a note that no derivative (a gradient is the multivariable version of a derivative) was supplied, so the "fminunc" used a method that doesn't require one.

```
>> lifeguard1
Warning: Gradient must be provided for trust-region algorithm; using quasi-newton algorithm instead.
> In fminunc (line 403)
  In lifeguard1 (line 10)

Local minimum found.

Optimization completed because the size of the gradient is less than
the default value of the function tolerance.

<stopping criteria details>

The minimum time is 40.1659 seconds and occurs at x = 87.2076 meters
>>
```

A more involved example implements the binary search method without specialized optimization functions. The code follows:

```
function lifeguard2

% This function solves the one variable lifeguard
problem using
% the binary search method.  It was intended to be
easy to read rather
% than efficient or elegant.  Feel free to refine it
:-)

clc % clear screen to keep things tidy

x(1)=0; % set the starting point
dx=20;  % set the initial step size
stop_flag=0; % define a flag that stops the initial
search
i=1; % initialize a counter
dx_min = 0.1; % Set exit criteria. Stop when dx <
dx_min

% Define the objective function using a MATLAB
anonymous function
T = @(x) 1/7*sqrt(50^2+x^2)+1/2*sqrt(50^2+(100-x)^2);

t(1)=T(x(1)); % calculate objectrive function at the
starting point
```

```
while stop_flag==0
    i=i+1;
    x(i)=x(i-1)+dx;
    t(i)=T(x(i));
    if t(i)>t(i-1)
        stop_flag=1;
    end
end

plot(x,t,'ob','MarkerSize',10) % Mark the steps with
blue circles
grid on
hold on   % Keep figure available for more content

% Start Binary Search

stop_flag=0; % Reset slop flag

x2=x(i-1);    t2=t(i-1);

while dx>dx_min
    dx=dx/2;
    x1=x2-dx;   t1=T(x1);
    x3=x2+dx;   t3=T(x3);
    tt=[t1 t2 t3];
    xx=[x1 x2 x3];
    [tmin,n]=min(tt);
    t2=tmin;
    x2=xx(n)
    plot(x2,t2,'ob','MarkerSize',10)

    disp(['Current estimate of x* = ' num2str(x2) '
meters'])
    disp(['Current estimate of T* = ' num2str(t2) '
seconds'])
    disp(['dx = ' num2str(dx)])
    disp([' '])

end

plot(x2,t2,'+r','MarkerSize',10) % Mark minimum with
red plus sign
hold off

return
```

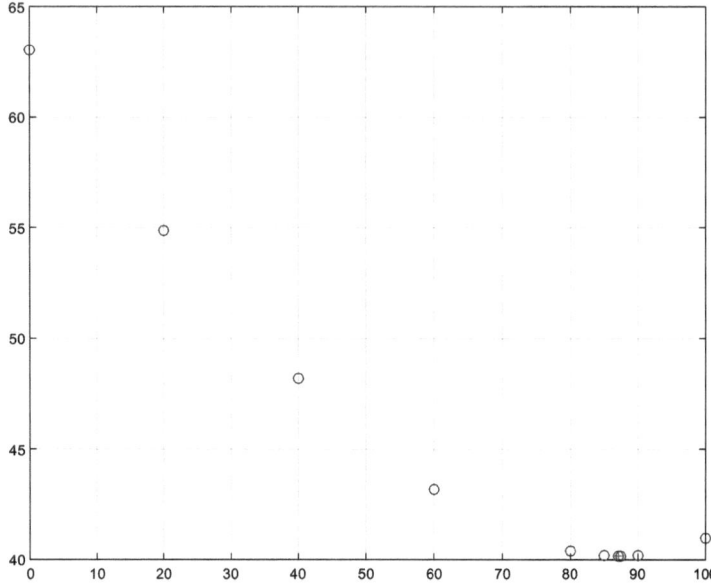

References

1. Vanderplaats G (2005) Numerical optimization techniques for engineering design, 4th edn. Vanderplaats Research and Development, Colorado Springs
2. Anderson J (2016) Fundamentals of aerodynamics, 6th edn. McGraw Hill, New York
3. Mathworks. MATLAB. Mathworks [Online]. Available: http://www.mathworks.com/products/matlab/. Accessed 12 June 2016
4. Press WH, Teukolsy SA, Vetterling WT, Flannery BP (2007) Numerical recipes: the art of scientific computing, 3rd edn. Cambridge University Press, New York
5. Jacob M (2001) Power electronics: principles and applications. Delmar Cengage, Independence
6. Hibbeler RC (2015) Engineering mechanics: dynamics, 14th edn. Pearson, Hoboken
7. Reddy J (2006) Theory and analysis of elastic plates and shells, 2nd edn. CRC Press, Boca Raton
8. Kroese D, Brereton T, Taimre T, Botev Z (2014) Why the Monte Carlo is so important today. WIREs Computational Statistics 6(6):386–392
9. Kroese D, Taimre T, Botev Z (2011) Handbook of Monte Carlo methods. Wiley, Hoboken
10. Knuth D (1998) The art of computer programming: vol 3: sorting and searching, 2nd edn. Addison-Wesley, Boston
11. Azen S (1966) Successive approximation by quadratic fitting as applied to optimization problems. Rand Corp, Santa Monica
12. Casson L (1995) Ships and seamanship in the ancient world. Johns Hopkins University Press, Baltimore
13. Anastasi P. Aegean sea floor yields clues to early Greek traders. New York Times, 2 January 1989
14. Elphick P (2006) Liberty: the ships that won the war. Naval Institute Press, Annapolis
15. Levinson M (2008) The box: how the shipping container made the world smaller and the world economy bigger. Princeton University Press, Princeton
16. Hennemand M (2012) Containers – talk about a revolution! ISO Focus+ 3(4):21–22
17. Faulkner FD (1963) Numerical methods for determining optimum ship routes. Navigation 10 (4):351–367
18. Hanssen GL, James RW (1960) Optimum ship routing. J Navig 13(3):253–272

Chapter 3
Minimum Principles: Optimization in the Fabric of the Universe

The most general law in nature is equity – the principle of balance and symmetry which guides the growth of forms along the lines of the greatest structural efficiency

-Sir Herbert Read

It may be tempting to think of optimization as a human invention, but nature is way ahead of us. There are many ways in which optimization-like concepts appear naturally in the world around us. In fact, optimization and minimum principles seem to be an integral part of creation on a very deep level [1].

An obvious example of optimization in the natural world is evolution, the survival and perpetuation of living things on the basis of which ones are most suited to their environments. Another is even more grand – general relativity. There are many others, though, that might be less familiar, including structural mechanics and the dynamics of moving bodies. It is clear that many of the principles used by engineers and physicists can be derived by showing that nature minimizes some quantity. Of these, evolution is perhaps the easiest to describe, so let's start there.

3.1 Evolution

Roughly speaking, natural selection is the process by which living things that are most well suited to their environments survive and reproduce. Those less suited to withstand the rigors of their environment are less likely to survive long enough to reproduce and their populations dwindle. The spark that makes evolution progress is that genetic differences spontaneously arise in living things. In this way, nature is continuously offering new and occasionally better solutions to the problem of survival.

It is quite plausible then to view the living world as an optimization problem in which the objective is to produce a population most suited to survive and to reproduce [2]. The objective function is reproductive success.

© Springer International Publishing AG, part of Springer Nature 2018
M. French, *Fundamentals of Optimization*,
https://doi.org/10.1007/978-3-319-76192-3_3

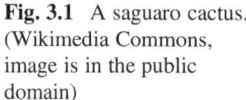

Fig. 3.1 A saguaro cactus. (Wikimedia Commons, image is in the public domain)

Over the last billion or so years, evolution has worked quite well. Life is found routinely in the most inhospitable environments; indeed, it is hard to find a place on Earth in which life doesn't exist. While every environment poses challenges, these challenges are easier to see in the extremes. A common example is the saguaro cactus that grows in the American southwest [3], an example of which is shown in Fig. 3.1.

It grows slowly and is able to extract water from an environment that has almost none. By collecting and storing water, it is able to survive extended droughts. The fact that it contains liquid water makes it a tempting target, but its liberal covering of sharp thorns keeps large animals from trying to eat it (though a number of small animals feed on it). All plants use photosynthesis to make energy. The desert is very sunny, so the saguaro doesn't need to waste energy growing leaves. Rather, the trunk has enough surface area to collect the light needed to grow. The trunk is green due to the chlorophyll that collects sunlight rather than brown because of a covering of bark.

In a real sense, the lowly cactus is optimally adapted to its environment. This does not mean that some other type of plant couldn't grow more successfully in that environment. The pace of evolution is basically a function of the rate of reproduction of the life-form. Saguaro cacti often live more than 100 years, so evolutionary changes generally come slowly. Conversely, the fruit flies so ubiquitous in biology labs have a reproductive cycle measured in weeks rather than decades and, so, are much better suited for studying genetic adaptations and variations. Evolutionary changes can sometimes be observed in the course of months.

One of the apparent rules of evolution is that everything is more complicated than it first appears. For example, the saguaro is very sensitive to extreme cold, and one or two nights of freezing temperatures are often enough to kill them. The result is that it is possible to map the distribution of freezing temperatures by mapping the distribution of live saguaro.

An inadvertent, man-made example of the power of evolution to optimize the characteristics of living things involves rabbits and Australia [4]. In the year 1859, a European settler, Thomas Austin, imported 24 rabbits, ostensibly so he could have something convenient to shoot at. With no natural predators and an abundance of plants on which to feed, they reproduced and became a scourge. It was clear that something else had to be introduced into the environment to prey on them. Rather than a predator that might also eat native species, the decision was made to introduce a virus, Myxomatosis, that affected only rabbits [5].

The virus had a high mortality rate, and the rabbit population fell rapidly. The mortality rate was not 100% though, and a small, resistant portion of the rabbit population survived infection. They reproduced and the population soon recovered, so Australia was again overrun with rabbits. This second population explosion, however, was comprised of rabbits resistant to the virus. In a clear way, this second group of rabbits is optimized for survival in an environment where they could be exposed to Myxomatosis. There are many other examples of evolution taking place before us, as in the case of numerous bacteria that are now resistant to common antibiotics.

These are but two of a huge number of possible examples of evolution in action. As far as we can tell, evolution, roughly as Darwin envisioned it, is the central fact of life on Earth. It is clearly an optimization process but one with a unique twist in that no analysis is necessary. Evolution doesn't know why some genetic variations work better than others, only that they do. There is no need for any kind of analysis since nature simply keeps those able to survive and coldly disposes of the rest. This is optimization by experiment rather than analysis.

The basic idea of evolution is easy to convey, and there is even a field of study within the field of optimization devoted to what are called genetic algorithms. These are optimization methods intended to mimic the behavior of evolution. A simple example may be enough to get the point across. Steve Jones, an eminent evolutionary biologist, has a practical application that is now well known [6].

He found himself working on a problem for a factory that made soap powder. This was done by forcing hot chemicals through a nozzle after which they condensed into a powder. The nozzle didn't work very well, and there was a concerted effort to improve it. The problem was that the underlying physics are very complex, and nobody understood the operation of the nozzle well enough to improve it. In what must have been an act of desperation, Jones, a biologist, was consulted and suggested what biologists know best, evolution.

Making nozzles was apparently not difficult, so ten random variations were introduced into the standard nozzle design. The best of that group was chosen, and there were ten random variations applied to that design – two generations so far. After 45 generations, they had a nozzle that worked much better than the original. The interesting thing is that nobody knew how that one worked either. With evolution, you don't need to know how an object or a creature works. You need only know that it does.

The other way minimum principles affect our lives is through the operation of the most basic physical processes. While perhaps less obviously part of our lives than evolution, minimum principles are even more universal and are independent of living things. They are majestic, universal, and mysterious. While we are able to describe how these principles act, nobody knows why they act. They constitute, in the opinion of the author, one of the greatest mysteries of the universe.

3.2 Minimum Energy Structures

A minimum principle that affects all of us every day concerns structures under load. The deformed shape of the structure is the one that minimizes its total energy. That's it. It doesn't matter whether the structure is a bridge, a building, or a bicycle; the principle is the same.

If, for instance, a chain is hung by its ends, it forms one very specific shape called a catenary [7]. If we assume the chain doesn't stretch under just gravitational forces, there is no strain energy, and only the potential energy is being minimized. No other shape of the infinitude of possible shapes gives the chain lower potential energy. Figure 3.2 shows a catenary shape produced by a chain hanging from its ends.

Since a chain has no bending stiffness, the links must be in tension only. Conversely, if the shape could be frozen and flipped vertically, the result would be an arch that is in compression only. Thus, an arch built in the shape of a catenary and loaded uniformly along its length can be made from a material incapable of resisting tensile loads, like stacked stones.

Figure 3.3 shows several examples of arches used in bridges and other large structures. Figure 3.3a shows a series of catenary arches in La Pedrera (also known as Casa Milà) designed by Antoni Gaudi. The arches are made from stacked blocks held together with mortar – materials unable to resist tensile loads. While the catenary is an optimal shape, it is not easy to build and can sometimes be approximated by simple shapes.

Figure 3.3b shows a section of Roman aqueduct near Pont du Gard, near Nîmes in France. This section is part of an approximately 50-km-long aqueduct and was probably built in the first century AD. The arches are semicircular with straight vertical extensions. They are apparently adequately designed, having stood for nearly two millennia of constant use and being constructed of hewn stones fit together without mortar. A much less serious but no less instructive example comes in the form of an arch made by stacking old CRT computer monitors as shown in Fig. 3.3c. They do not appear to be fixed together with glue or bolts.

Fig. 3.2 A hanging chain showing the shape of a catenary

Rather, the perpendicular force across the faces of the monitors holds them together, just like in a stone arch.

Minimum energy is not the only way to calculate the catenary shape. It can also result from a simple balance of forces using concepts familiar to any freshman engineering student. It is one of the intriguing aspects of minimum principles that they can be used to derive physical laws that are also described by more direct methods; seemingly unrelated arguments can be used to arrive at the same conclusion.

On a more accessible level, minimum principles are very nicely demonstrated by the simple soap bubble [8]. The film formed by dipping a wire loop in a bucket of soapy water is a surface that minimizes the surface energy. This is equivalent to minimizing the surface area of the film. The math describing this phenomenon is interesting but detailed and will be discussed later. For now, it is enough to know that soap films are a fun way of experimentally solving a class of complex math problems.

Figure 3.4 shows a soap film joining two circular wire loops. Initially, one might guess that the soap would form a straight cylinder. However, a ring with a pronounced waist has the lowest possible surface area of any shape that could join the two loops. This has been proven mathematically; again minimum principles are shown to underlie physical phenomena.

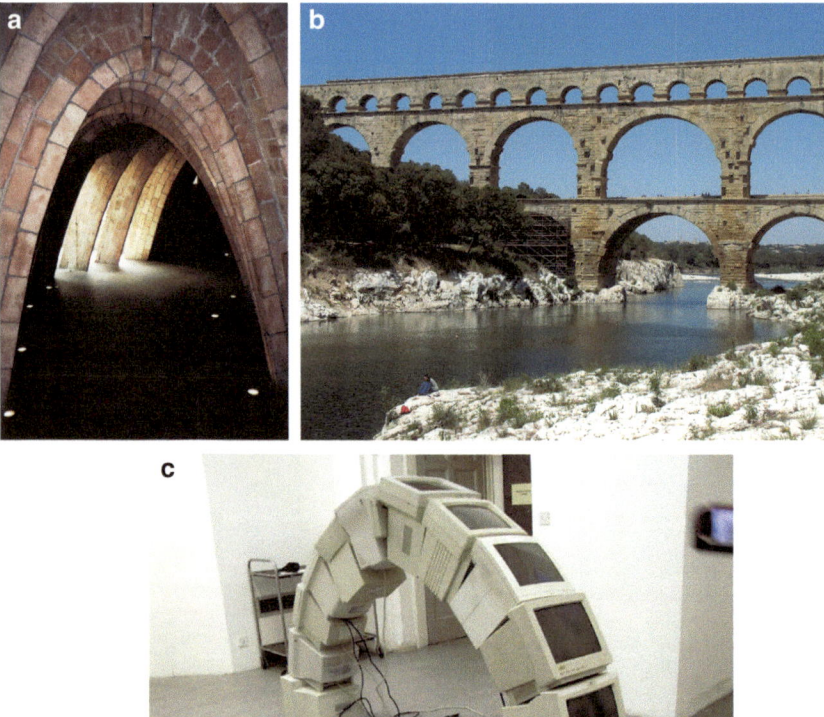

Fig. 3.3 (**a**) Catenary arches under the roof of Gaudí's Casa Milà, Barcelona, Spain (Wikipedia Commons, image is in the public domain). (**b**) Roman aqueduct in Pont du Gard, France (Wikipedia Commons, image is in the public domain). (**c**) An arch made from computer monitors. (Image available on many Internet sites, original source unknown)

Let's see how a simple mathematical model compares to the soap film in Fig. 3.4. Rather than trying to duplicate the curve, consider a simpler model that uses two truncated cones as shown in Fig. 3.5

The goal is to find the surface with the smallest possible surface area by changing the radius, r, where the two cones join. Note that the shape is symmetric about the plane of intersection.

The area of the curved portion of the two truncated cones is

$$A = 2\pi(R + r)\sqrt{h^2 + (R - r)^2} \qquad (3.1)$$

Fig. 3.4 A soap film joining two wire loops. (Wikimedia Commons, image is in the public domain)

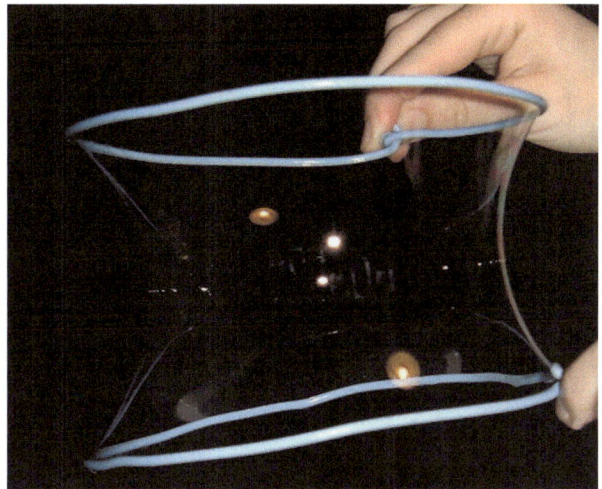

Fig. 3.5 Geometry of simple soap film model

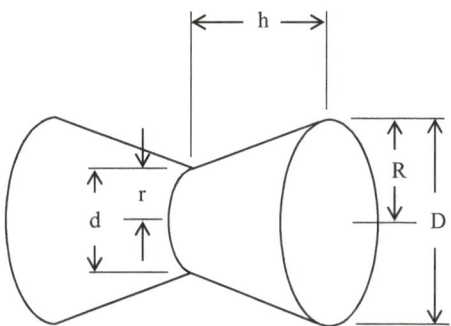

The result, shown in Fig. 3.6, gives the resulting center radius as 42.678 mm or a diameter of 85.356 mm. This matches roughly the dimensions shown in Fig. 3.4.

As a check, it is not a problem to write out the derivative of the objective function, dA/dr, and substitute $r*$ for r. Using the numerical value $r* = 42.678$ gives $dA/dr = 0.00546$, close to zero.

Of course, a more sophisticated model should give better correlation with Fig. 3.4 and should also give a lower surface area than the previous model. Since the soap film at the halfway point between the two hoops is parallel to the center line, let's make a slight change to the previous math model as shown in Fig. 3.7. The addition of a straight center section may allow a slightly lower total surface area since it would appear to be a slightly better approximation to the actual surface shown in Fig. 3.4.

Figure 3.8 shows the calculations for the improved model. The resulting surface area is indeed slightly smaller than the first, even though the final diameter is slightly larger.

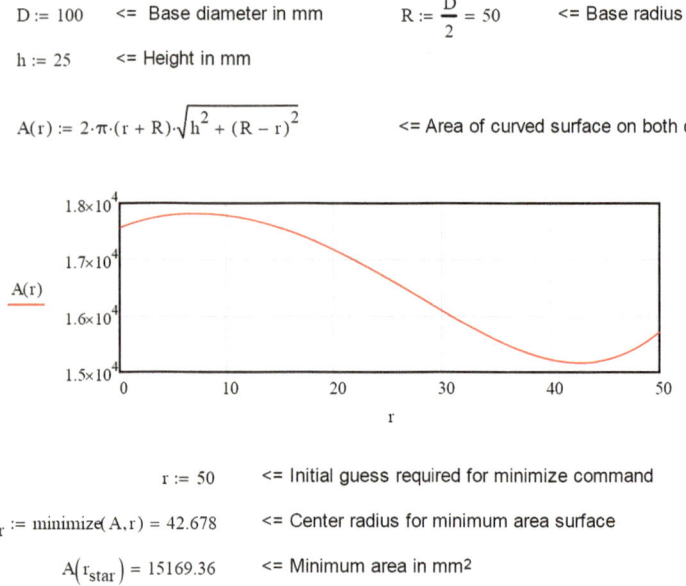

$D := 100$ <= Base diameter in mm $R := \dfrac{D}{2} = 50$ <= Base radius in mm

$h := 25$ <= Height in mm

$A(r) := 2 \cdot \pi \cdot (r + R) \cdot \sqrt{h^2 + (R - r)^2}$ <= Area of curved surface on both cones

$r := 50$ <= Initial guess required for minimize command

$r_{star} := \text{minimize}(A, r) = 42.678$ <= Center radius for minimum area surface

$A(r_{star}) = 15169.36$ <= Minimum area in mm2

Fig. 3.6 Simple soap film model

Fig. 3.7 Revised math model for soap film

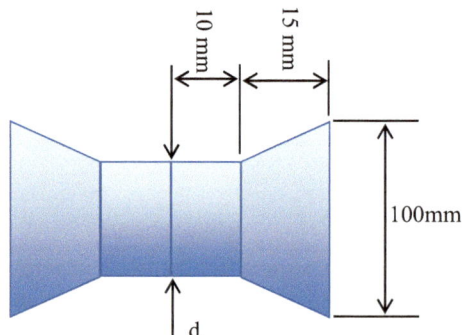

Another way to address the problem is to assume a smooth curve that connects the end points of the soap film and calculate the area of the surface that results when the curve is rotated about the central axis. Since the problem is symmetric about the y axis, it makes sense to use a function that is also symmetric. In this chapter, we are limiting ourselves to problems with a single variable, so let's use a second-order polynomial

$$y = ax^2 + bx + c \tag{3.2}$$

There are two required boundary conditions, so two of these constants will be defined by the third. Since the curve needs to be symmetric, $b = 0$. Next, we can apply the boundary condition $y(-25) = y(25) = 50$ to find that $c = 50 - 625a$. The result is a function of a single variable, a (Fig. 3.9).

$D := 100$ $h_1 := 10$ $h_2 := 15$ $R := \dfrac{D}{2} = 50$

$A_1(r) := 2 \cdot \pi \cdot r \cdot h_1$ <= Area of cylindrical portion

$A_2(r) := \pi \cdot (R + r) \cdot \sqrt{h_2{}^2 + (R - r)^2}$ <= Area of conical portion

$A(r) := 2 \cdot \left(A_1(r) + A_2(r) \right)$ <= Total area

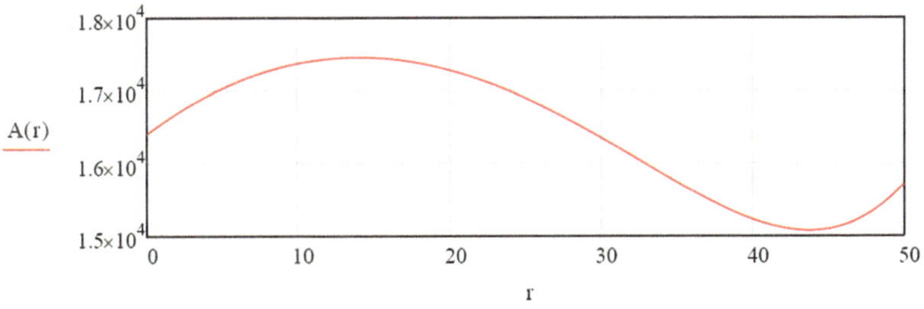

$r := 50$ <= Initial guess required for minimize command

Given $r > 0$ <= Require that radius be positive

$r_{star} := \text{Minimize}(A, r) = 43.703$ <= Center radius for minimum area surface

$A(r_{star}) = 15069.8$ <= Minimum area in mm²

Fig. 3.8 Revised soap film model

$$y = ax^2 - 625a + 50 \tag{3.3}$$

To find the resulting surface area, we can imagine the axisymmetric shape being cut into sections. Figure 3.10 shows the shape of a single section.

The surface area of this tapered ring is $A = \pi D ds = 2\pi y ds$. Note that D is the average diameter. This is closely related to the surface area of a truncated cone used in Eq. (3.1).

For convenience, we need to express ds as a function of x and y. This is just a straightforward application of the Pythagorean theorem.

$$ds^2 = dx^2 + dy^2 \quad \text{or} \quad \left(\frac{ds}{dx} \right)^2 = 1 + \left(\frac{dy}{dx} \right)^2 \tag{3.4}$$

So it becomes clear that

$$ds = \sqrt{1 + \left(\frac{dy}{dx} \right)^2}\, dx \tag{3.5}$$

The area of a single ring is then

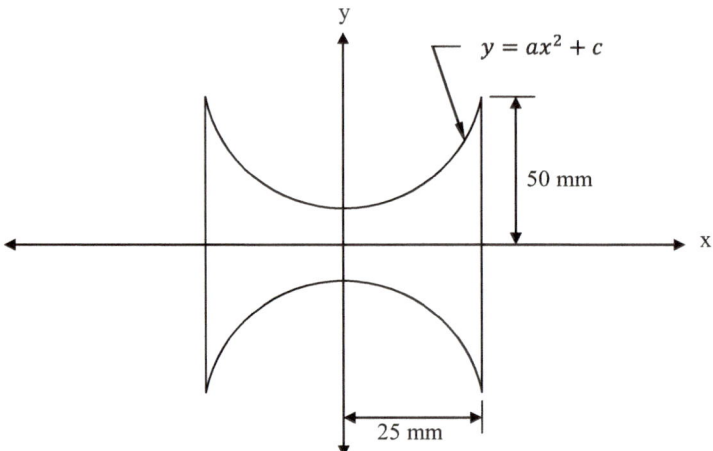

Fig. 3.9 Soap film modeled with a polynomial

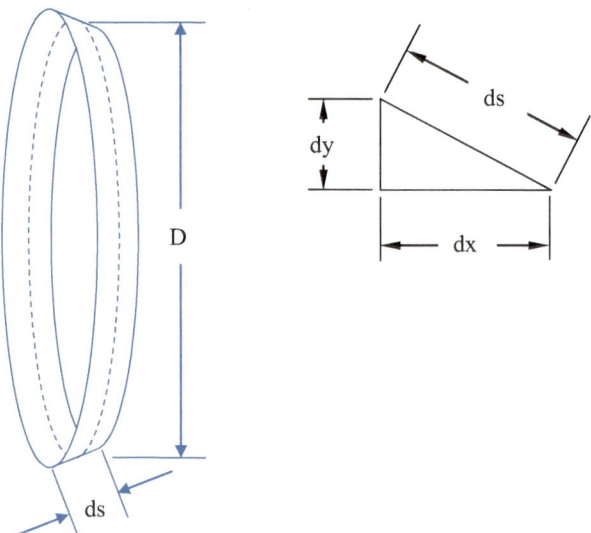

Fig. 3.10 Geometry of a section of the assumed shape of the soap film

$$dA = 2\pi y\sqrt{1 + \left(\frac{dy}{dx}\right)^2}\,dx \qquad (3.6)$$

The total area of the surface is just the sum of the areas of rings that make up the shape. If the width of the rings becomes arbitrarily small, the total area is the integral from one end of the shape to the other along the x axis.

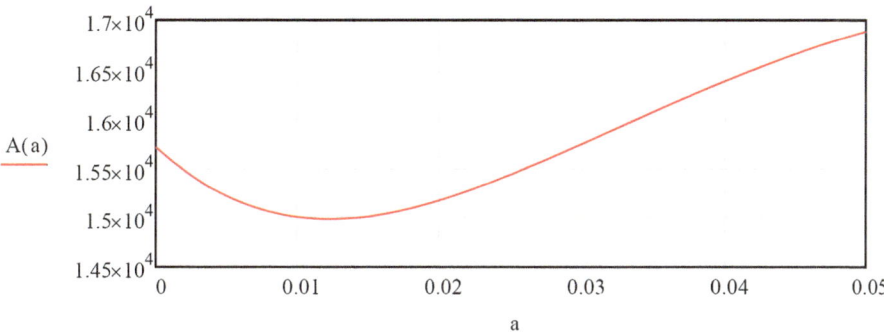

Fig. 3.11 Area as a function of polynomial parameter, a

$a := 0$ $\Leftarrow$ Initial guess - needed by minimization function

$a_{star} := \text{minimize}(A, a) = 0.0122$ $\Leftarrow$ Find value of a to minimize A

$A(a_{star}) = 14979.646$ $\Leftarrow$ Minimum area using assumed shape

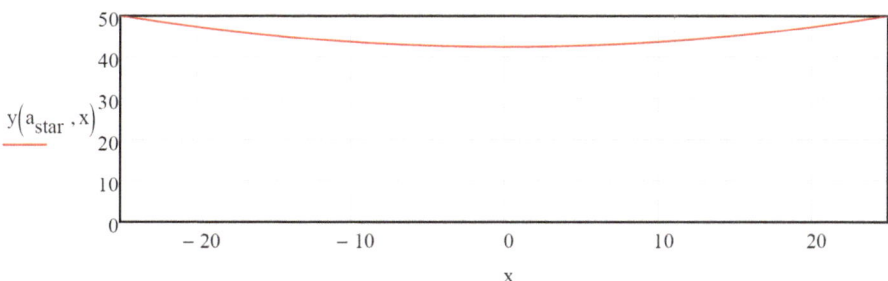

Fig. 3.12 Shape of minimum area surface using assumed function

$$A = \int_{-25}^{25} 2\pi y \sqrt{1 + \left(\frac{dy}{dx}\right)^2}\, dx = \int_{0}^{25} 4\pi y \sqrt{1 + \left(\frac{dy}{dx}\right)^2}\, dx \qquad (3.7)$$

Filling in the complete expression for y gives

$$A = \int_{0}^{25} 4\pi \left(ax^2 - 625a + 50\right) \sqrt{1 + (2ax)^2}\, dx \qquad (3.8)$$

Now, A is a function of a single variable, a, and it's easy to plot, as seen in Fig. 3.11.

Area is thus a minimum when $a = 0.0122$. Figure 3.12 shows the calculation of the minimum area and the corresponding shape. Note that minimum area is slightly smaller than those found using the previous models, and the center radius is near that predicted by the previous two models. While not strictly a check, it does lend confidence.

It is important to note, even though this is a smooth shape, it is still only approximate since it is based on a second-order polynomial. The choice of the approximating function was based on convenience, and there is no reason to think the shape of the minimum surface is actually a second-order polynomial.

In the soap film problems presented here, we used a single design variable to approximate a minimum principle in nature. Note that these examples have only a single design variable because that's all we've covered so far. With more design variables, we could get closer to the exact solution, which is itself continuous and belongs to a field called the calculus of variations. It is very interesting but far beyond the limits of the current discussion.

3.3 Optics: Fermat's Principle and Snell's Laws

There are minimum principles at the heart of many physical laws, and optics is no exception. In 1662, Pierre de Fermat made a conceptual leap by proposing that light always travels between two points along the path that minimizes time [9]. It was later shown that the actual mathematical requirement is slightly more nuanced. However, for many simple problems, Fermat's principle can be applied without complication. One of these is the path taken by light being reflected from a mirror as shown in Fig. 3.13.

Snell's law of reflection says that $a/L_1 = a/L_2$ (angle of incidence equals angle of reflection), and Fermat's principle says that x will take the value that minimizes the time required for the light to travel in a straight line in a uniform medium (like air in a room). Transit time is $T = L/c$ where c is the speed of light and L is the distance traveled. The speed of light is approximately 3×10^8 m/s. The time required for the light to get from P_1 to P_2 is

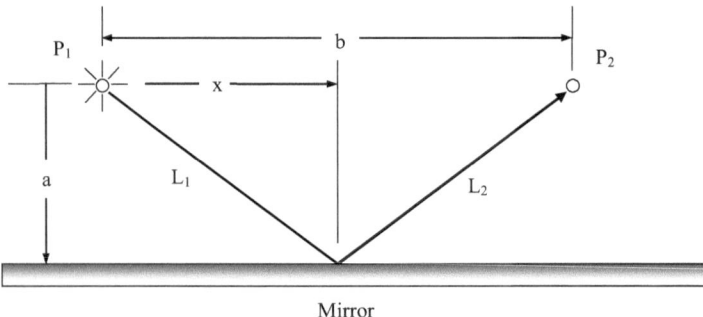

Fig. 3.13 Path of light from source to sensor

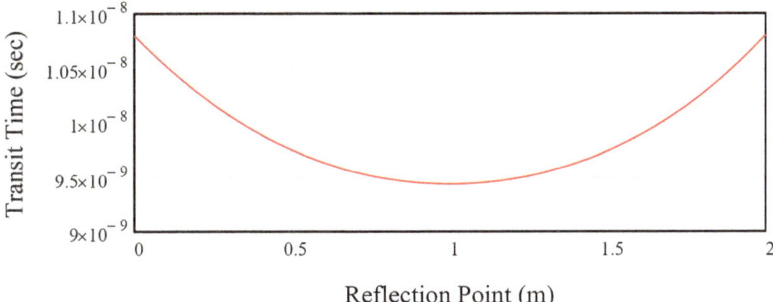

Fig. 3.14 Transit time as a function of reflection point

$$T = \frac{L_1}{c} + \frac{L_2}{c} = \frac{1}{c}\left[\sqrt{a^2 + x^2} + \sqrt{a^2 + (b - x)^2}\right] \tag{3.9}$$

Say that $a = 1$ m and $b = 2$ m. Intuition and Snell's law both suggest that $x^* = 1$ m. A plot of transit time as a function of x confirms this (Fig. 3.14).

Again, it is not difficult to find x^* by finding the root of dT/dx.

$$\frac{dT}{dx} = \frac{1}{c}\left[\frac{x}{\sqrt{x^2 + a^2}} + \frac{x - b}{\sqrt{a^2 + b^2 - 2bx + x^2}}\right] = 0 \tag{3.10}$$

Solving Eq. (3.10) numerically results in $x^* = 1$. To solve the problem analytically, this equation can be slightly modified.

$$\frac{dT}{dx} = \frac{1}{c}\left[\frac{x\sqrt{a^2 + b^2 - 2bx + x^2} + (x - b)\sqrt{a^2 + x^2}}{\sqrt{a^2 + x^2}\sqrt{a^2 + b^2 - 2bx + x^2}}\right] = 0 \tag{3.11}$$

Only the numerator need be zero, so solving this equation is equivalent to solving

$$x\sqrt{a^2 + b^2 - 2bx + x^2} + (x - b)\sqrt{a^2 + x^2} = 0 \tag{3.12}$$

This equation is satisfied if $x = b/2$. Thus, the solution using Fermat's principle agrees with Snell's law of reflection. That means that a minimum principle underlies one of the most basic laws of optics.

A ray of light moving from one medium to another one bends as long as the incident angle isn't 90°. This is how lenses and prisms work. The key is the fact that the speed of light varies depending on the medium through which it is moving. The speed of light in a vacuum is well known to be 3×10^8 m/s (186,000 miles/s) and is denoted by c, after *celeritas*, the Latin word for velocity. However, this speed decreases in dense media. For example, the speed of light in water is about 3/4 the speed in a vacuum. A quantity called the refractive index, η, is the ratio of c to the speed of light in the medium. The refractive index of water is, thus, about 4/3. It is worth noting that the index of refraction is also a weak function of wavelength – this

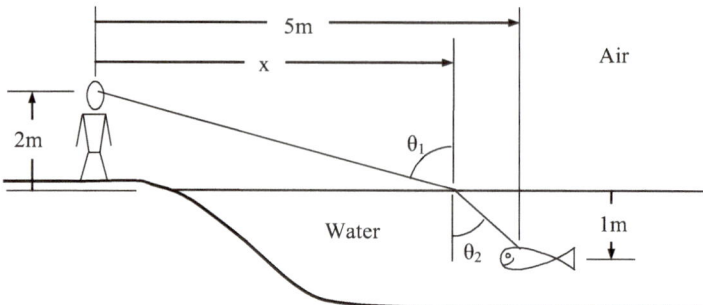

Fig. 3.15 Line of sight through water

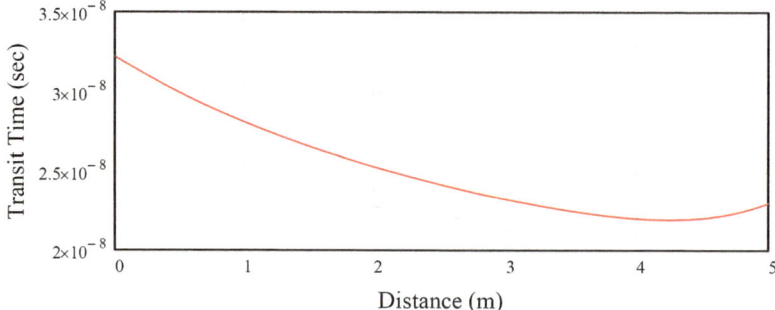

Fig. 3.16 Time as a function of distance

is why a prism can decompose white light into a rainbow – but this complication
isn't necessary here.

Fermat's principle also applies to problems of refraction. Consider the familiar
example of looking at a fish below the surface of a pond as shown in Fig. 3.15.

The time required for a light ray to go from the fish to the observer is the sum of
the time required for the two segments. Note that Eq. (3.13) is very similar to that for
the lifeguard problem. Mathematically speaking, the two equations are almost
identical.

$$T = \frac{1}{c}\sqrt{2^2 + x^2} + \frac{\eta_w}{c}\sqrt{1^2 + (5 - x)^2} \tag{3.13}$$

where $\eta_w = 4/3$. Figure 3.16 shows a plot of time as a function of x. It is clear that the
minimum time is around 4.1 m. Solving numerically for the minimum gives
$x^* = 4.088$.

Snell's law of refraction states $\eta_1\sin\theta_1 = \eta_2\sin\theta_2$. For this problem, $\theta_1 = 63.933°$
and $\theta_2 = 42.365°$. As expected, $\sin(63.933°) = 4/3\sin(42.365°) = 0.898$, so the
solution found using Fermat's principle also satisfies Snell's law of refraction.
Again, a minimum principle underlies a basic law of optics.

3.4   General Relativity

Minimum principles seem to be woven into the very fabric of the universe [10]. Einstein's general theory of relativity, first proposed in 1915, is one of the most thoroughly verified theories in history and one of the most revolutionary. More recently, a satellite experiment called Gravity Probe B used perfectly spherical flywheels to demonstrate a very subtle prediction of general relatively called frame dragging [11]. The effect is a minute one and was predicted to be observable by a very small procession in the rotational axes of the gyroscopes. Indeed, the axis of rotation of the gyroscopes advanced much less than a degree over many months and did so in agreement with theoretical predictions. As I write this, gravity wave astronomy is emerging as a field within physics after the detection of gravity waves as predicted by general relativity.

It is popularly understood that Einstein replaced the Newtonian ideas of absolute space and absolute time with a construct in which they are affected by relative velocity and the presence of large masses. The idea that relativity connects space and time into a single four-dimensional entity is also familiar. Less well known is the central idea that moving bodies take a path through space-time, called a geodesic. The math is daunting, and the results are profound, but one result is that light follows the straightest path through four-dimensional space-time that has been curved by the

Fig. 3.17 Hoag's object, a ring formed by gravitational lensing. (Wikimedia Commons, image is in the public domain)

presences of massive bodies [12]. The path of light moving through space-time is thus dictated by a very deep minimum principle.

A directly observable effect of special relativity is called gravitational lensing. When viewed from the side, a spiral galaxy is roughly the shape of a lens used to focus light. The mass of the galaxy is sufficient to warp space-time so that light passing it is bent and focused on a distant point. Figure 3.17 shows a ring produced by light being bent as it passes a galaxy.

References

1. Born M (1939) Cause, purpose and economy of natural laws – minimum principles in physics. Nature 143(3618):357–361
2. Smith JM (1978) Optimization theory in evolution. Ann Rev Ecol Syst 9:31–56
3. Bowers R, Bowers N (2008) Cactus of Arizona field guide. Adventure Publications, Cambridge
4. Bryson B (2001) In a sunburned country. Broadway Books, New York
5. Bartrip PW (2008) Myxomatosis: a history of pest control and the rabbit. I.B. Tauris, London
6. Jones S (2000) Almost like a whale: the 'origin of species' updated. Doubleday, New York
7. Gil JB (2005) The catenary (almost) everywhere. Boletın de la Asociacion Matematica Venezola 12(2):251–258
8. Isenberg C (1992) The science of soap films and soap bubbles. Dover, Mineola
9. Feyman RP, Leighton RB (2005) The Feynman lectures, vol 1. Addison Wesley, Boston
10. Wolfson R (2003) Simply Einstein: relativity demystified. W. W. Norton & Company, New York
11. Everitt C et al (2011) Gravity probe B: final results of a space experiment to test general relativity. Phys Rev Lett 106(22):221101-1–221101-5
12. Haugan M (2017) Personal conversation

Chapter 4
Problems with More than One Variable

Continuous improvement is better than delayed perfection

-Mark Twain

Until now, we have been focused on problems with only a single design variable. However, few useful problems have just one design variable, so we need to move on to larger problems. Fortunately, the number of design variables only needs to be bigger than one in order to learn about general, multivariable solution methods. A method that works on two variables will also work on more than two variables without conceptual changes. We will focus on two-variable problems since they have the advantage in that it is still possible to plot design space.

It might help to see how the methods in this chapter progress from simplest to the most sophisticated. The methods are:

- Direct Solution – set a gradient (multivariable version of a derivative) equal to zero and solve the resulting equations.
- Monte Carlo – find minimum objective function value at a group of randomly selected points in design space.
- Marching Grid – the simplest method that uses a directional search.
- Steepest Descent – the simplest search method that uses derivatives.
- Conjugate Gradient – a more refined method that uses derivatives.
- Newton's Method – a method that uses both first and second derivatives.
- Quasi-Newton's Method – a more refined method that uses first and second derivatives.

4.1 Two-Variable Lifeguard Problem

Let's start with a variation on what now must by now be a familiar problem, the lifeguard problem. Previously, we assumed a speed of 7 m/s across the sand and 2 m/s in the water. However, we all know that it is possible to run slowly through shallow

© Springer International Publishing AG, part of Springer Nature 2018
M. French, *Fundamentals of Optimization*,
https://doi.org/10.1007/978-3-319-76192-3_4

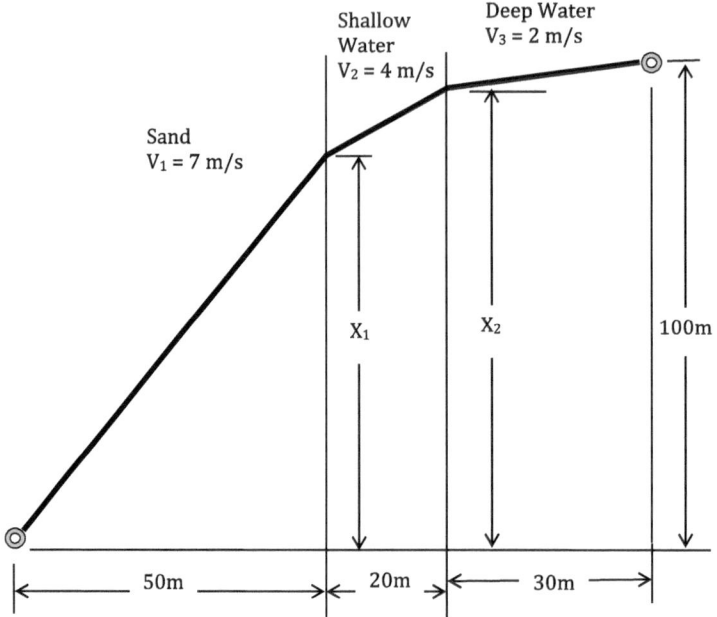

Fig. 4.1 Two-variable lifeguard problem

water before it gets deep enough that one must swim. Let's suppose that the lifeguard can run at 4 m/s for the first 20 m through the water as shown in Fig. 4.1.

Again, we assume that velocity is constant in each section and the total time is just the sum of the time taken to traverse each section.

$$T = \frac{L_1}{V_1} + \frac{L_2}{V_2} + \frac{L_3}{V_3} \tag{4.1}$$

Filling in distances and velocities gives

$$T = \frac{1}{7}\sqrt{50^2 + x_1^2} + \frac{1}{4}\sqrt{20^2 + (x_2 - x_1)^2} + \frac{1}{2}\sqrt{30^2 + (100 - x_2)^2} \tag{4.2}$$

Since there are two design variables, design space has three dimensions, the maximum number that can easily be represented on a two-dimensional sheet of paper or computer screen. Figure 4.2 shows a plot of design space. Each contour line joins points with the same values of T, so the closer the contour lines, the steeper the slope. The "+" indicates the location at which T is a minimum (found using the "minimize" function in Mathcad).

From the plot of design space, it appears that the minimum time is approximately at the point $x_1^* = 81$ m and $x_2^* = 92$ m.

Were this a single-variable problem, it would be possible to set the derivative of the objective function equal to zero and find the minimum. Fortunately there is a

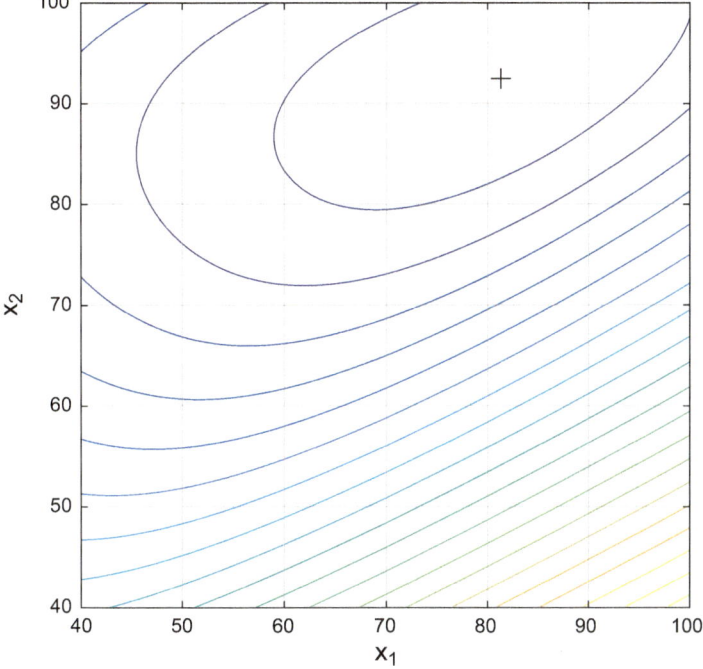

Fig. 4.2 Design space for the two-variable lifeguard problem

multidimensional version of the derivative called the gradient (see Appendix A) that lets us do the same thing with functions of multiple variables. The gradient of a function is a vector where each component is a partial derivative of the objective function [1]. In the case of this two-variable problem, the gradient is

$$\nabla T = \left\{ \begin{array}{c} \dfrac{\partial T}{\partial x_1} \\[2mm] \dfrac{\partial T}{\partial x_2} \end{array} \right\} \tag{4.3}$$

The elements of the gradient are just the slopes in each direction. Setting each of them equal to zero gives a set of equations to be solved. The solution to this set of equations is either a minimum or a maximum of the objective function.

$$\begin{aligned} \frac{\partial T}{\partial x_1} &= 0 \\[2mm] \frac{\partial T}{\partial x_2} &= 0 \end{aligned} \tag{4.4}$$

This is a bit cumbersome to do by hand, but not difficult in some convenient calculation software, as shown in Fig. 4.3. It is clear that setting the gradient terms equal to zero and solving gives the same result as using the minimize function in Mathcad.

Setting the gradient equal to zero and solving for the design variables, x_1 and x_2, is exactly equivalent to using the minimization function in a piece of computational software as shown in Fig. 4.4.

Analytical Solution of Two Variable Lifeguard Problem

$$T(x1,x2) := \frac{1}{7}\sqrt{50^2 + x1^2} + \frac{1}{4}\sqrt{20^2 + (x2 - x1)^2} + \frac{1}{2}\sqrt{30^2 + (100 - x2)^2}$$

Find the gradient of the objective function

$$S(x1,x2) := -\begin{pmatrix} \dfrac{d}{dx1}T(x1,x2) \\[2mm] \dfrac{d}{dx2}T(x1,x2) \end{pmatrix} \rightarrow \begin{bmatrix} -\dfrac{x1}{7\sqrt{x1^2 + 2500}} - \dfrac{2\cdot x1 - 2\cdot x2}{8\cdot\sqrt{(x1 - x2)^2 + 400}} \\[4mm] \dfrac{2\cdot x1 - 2\cdot x2}{8\cdot\sqrt{(x1 - x2)^2 + 400}} - \dfrac{2\cdot x2 - 200}{4\cdot\sqrt{(x2 - 100)^2 + 900}} \end{bmatrix}$$

x1 := 50   x2 := 50   <= Initial guesses for solver

Given $S(x1,x2) = \begin{pmatrix} 0 \\ 0 \end{pmatrix}$ $\begin{pmatrix} x1_{star} \\ x2_{star} \end{pmatrix} := find(x1,x2) = \begin{pmatrix} 81.326 \\ 92.472 \end{pmatrix}$ <= Minimum Point

$T\left(x1_{star}, x2_{star}\right) = 34.827$ <= Minimum Time

Fig. 4.3 Mathcad solution of two-variable lifeguard problem

Solution of Two Variable Lifeguard Problem Using Minimize Function

$$T(x1,x2) := \frac{1}{7}\sqrt{50^2 + x1^2} + \frac{1}{4}\sqrt{20^2 + (x2 - x1)^2} + \frac{1}{2}\sqrt{30^2 + (100 - x2)^2}$$

x1 := 50   x2 := 50   <= Initial guesses for minmizer

$\begin{pmatrix} x1_{star} \\ x2_{star} \end{pmatrix} := minimize(T,x1,x2) = \begin{pmatrix} 81.326 \\ 92.472 \end{pmatrix}$ <= Minimum Point

$T\left(x1_{star}, x2_{star}\right) = 34.827$ <= Minimum Time

Fig. 4.4 Solving the two-variable lifeguard problem using minimization

4.2 Least Squares Curve Fitting

Perhaps the most familiar multivariable optimization problem is curve fitting [2], also called regression. It is built into a popular commercial spreadsheet software and is routinely used as a way of identifying trends in measured data. The objective is to identify a function that passes as closely as possible through all the data points to minimize error. The total error is the objective function.

The first step is to define error. Error can be defined as vertical distance between the fitting function and the known data points as shown in Fig. 4.5.

Total error, E, might be just the sum of the individual errors. Note that, in this figure, one of the errors is negative, so it might be possible for E to be zero even though none of the individual error terms were zero. The obvious solution is to square each of the error terms so that total error is zero only when each of the squared error terms is zero. This also means that the total error is never negative.

$$E = \Delta_1^2 + \Delta_2^2 + \cdots + \Delta_n^2 \tag{4.5}$$

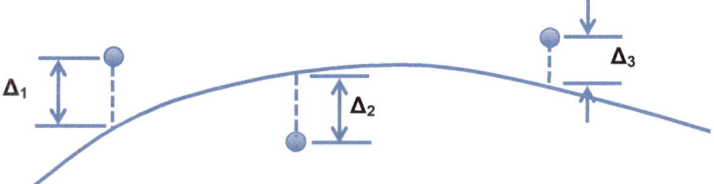

Fig. 4.5 Error in curve fitting

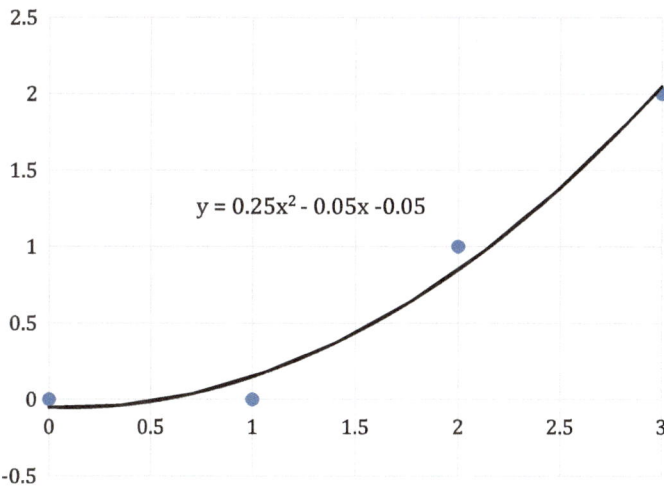

$$y = 0.25x^2 - 0.05x - 0.05$$

Fig. 4.6 A least squares curve fit

For a given set of data, the curve that minimizes E is the least squares curve fit. Figure 4.6 shows a least squares curve fit of four data points done using a spreadsheet program (Excel). The fitting function is a parabola but could be any appropriate function.

Let $P(x)$ be the fitting function

$$P(x) = ax^2 + bx + c \tag{4.6}$$

Then the squared error function is

$$E = (y_1 - p(x_1))^2 + (y_2 - p(x_2))^2 + (y_3 - p(x_3))^2 + (y_4 - p(x_4))^2 \tag{4.7}$$

This is an optimization problem in which the three design variables are a, b, and c and the objective function is E. Remember that x and y represent known data points, not variables as they often do in other equations. We need to find the values of a, b, and c that minimize E. Figure 4.7 shows the calculation in Mathcad. It's not a surprise that this solution matches the one from Excel.

$$x := \begin{pmatrix} 0 \\ 1 \\ 2 \\ 3 \end{pmatrix} \qquad y := \begin{pmatrix} 0 \\ 0 \\ 1 \\ 2 \end{pmatrix} \qquad \text{<= Assign the x and y vectors}$$

$$p(a,b,c,x) := a \cdot x^2 + b \cdot x + c \qquad \text{<= Define the fitting function}$$

$$E(a,b,c) := \sum_{i=1}^{4} \left(y_i - p\left(a,b,c,x_i\right) \right)^2 \qquad \text{<= Define squared error}$$

$$a := 0 \qquad b := 0 \qquad c := 0 \qquad \text{<= Initial guesses for unknown parameters}$$

$$\begin{pmatrix} a \\ b \\ c \end{pmatrix} := \text{Minimize}(E,a,b,c) = \begin{pmatrix} 0.25 \\ -0.05 \\ -0.05 \end{pmatrix} \qquad \text{<= Identify a,b and c by minimizing squared error}$$

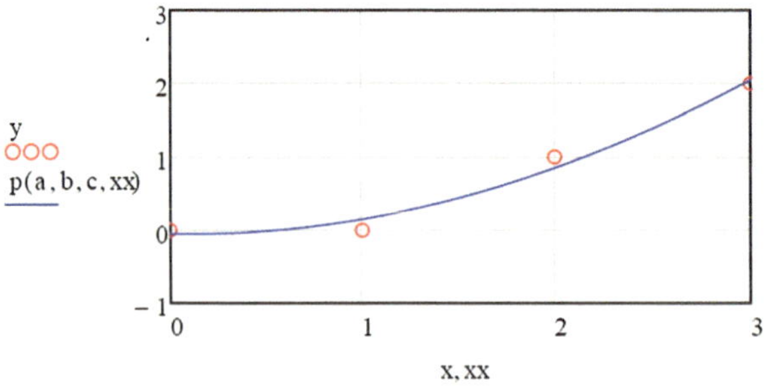

Fig. 4.7 Least squares curve fit calculation

Fig. 4.8 Two-variable soap
film problem

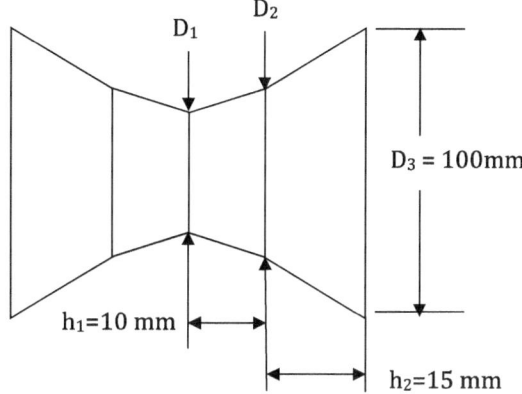

4.3   Two-Variable Soap Film Problem

For another example let's revisit the soap film problem and give it two design
variables rather than one as shown in Fig. 4.8.

By adding the second design variable, the approximate shape of the film should
be closer to the exact solution (within the limits of our assumed geometry). In the
language of optimization, design space now has an additional dimension, so there is
much more room through which to look for a minimum solution.

The surface area of the smaller tapered ring is

$$A_1 = \pi(R_1 + R_2)\sqrt{h_1^2 + (R_2 - R_1)^2} \qquad (4.8)$$

And the surface area of the larger tapered ring is

$$A_2 = \pi(R_2 + R_3)\sqrt{h_2^2 + (R_3 - R_2)^2} \qquad (4.9)$$

Since the shape is symmetric, left to right, the total area is simply $2(A_1 + A_2)$. The
calculation is shown in Fig. 4.9. Note that the minimum area is smaller than that
predicted by the single-variable model in the previous chapter.

It's also possible to use an assumed curve as in the example from the previous
chapter. Since we are now considering problems with more than one design variable,
we might assume a quartic polynomial to describe the curve. After applying the two
end conditions, there will still be two design variables.

$$y = ax^4 + bx^2 + c \qquad (4.10)$$

As before, only the symmetric terms in the polynomial are used. Since y
$(25) = 50$, it is possible to solve for one of the unknown parameters in terms of
the two others.

Two Variable Soap Film Problem

$R_1 := 30$ $R_2 := 30$ $h_1 := 12.5$ $h_2 := 12.5$ $R_3 := 50$

$$A(R_1, R_2) := \pi \cdot (R_1 + R_2) \cdot \sqrt{h_1^2 + (R_2 - R_1)^2} + \pi \cdot (R_2 + R_3) \cdot \sqrt{h_2^2 + (R_3 - R_2)^2}$$

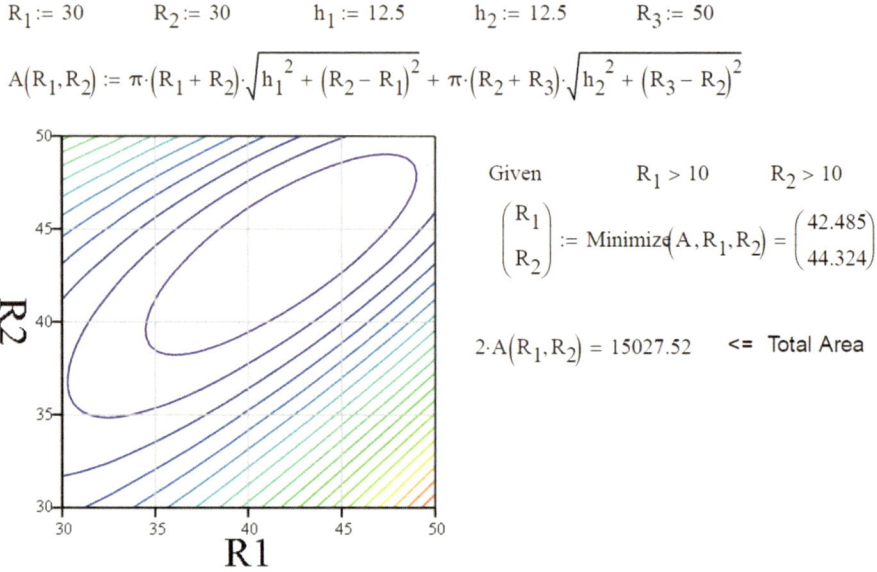

Given $R_1 > 10$ $R_2 > 10$

$$\begin{pmatrix} R_1 \\ R_2 \end{pmatrix} := \text{Minimize}(A, R_1, R_2) = \begin{pmatrix} 42.485 \\ 44.324 \end{pmatrix}$$

$2 \cdot A(R_1, R_2) = 15027.52$ <= Total Area

Fig. 4.9 Calculation for two-variable soap film problem

$$y(25) = 25^4 a + 25^2 b + c = 390625 a + 625 b + c = 50 \tag{4.11}$$

So the assumed function reduces to

$$y = ax^4 + bx^2 + (50 - 390625a - 625b) \tag{4.12}$$

where a and b, the unknown parameters, are the design variables. As before, the area of the surface of rotation is the objective function. Figure 4.10 shows the calculations in Mathcad. Note that a small modification has been used to ensure that the answer is realistic. There is nothing about the objective function, A, that requires the assumed function, y, to be positive everywhere. Rather than write a constraint that requires y to be positive everywhere, there are simply two constraints that require y to be positive at $x = 0$ and $x = 10$, on the assumption that any curve that is positive at these two points will be positive throughout the region of interest. This is, admittedly, a blunt solution, but it works well enough; the calculated solution is indeed positive everywhere.

Note that the quartic solution is almost identical to the quadratic solution from the previous chapter. The parameter, a, is very close to zero, so the quartic term isn't needed.

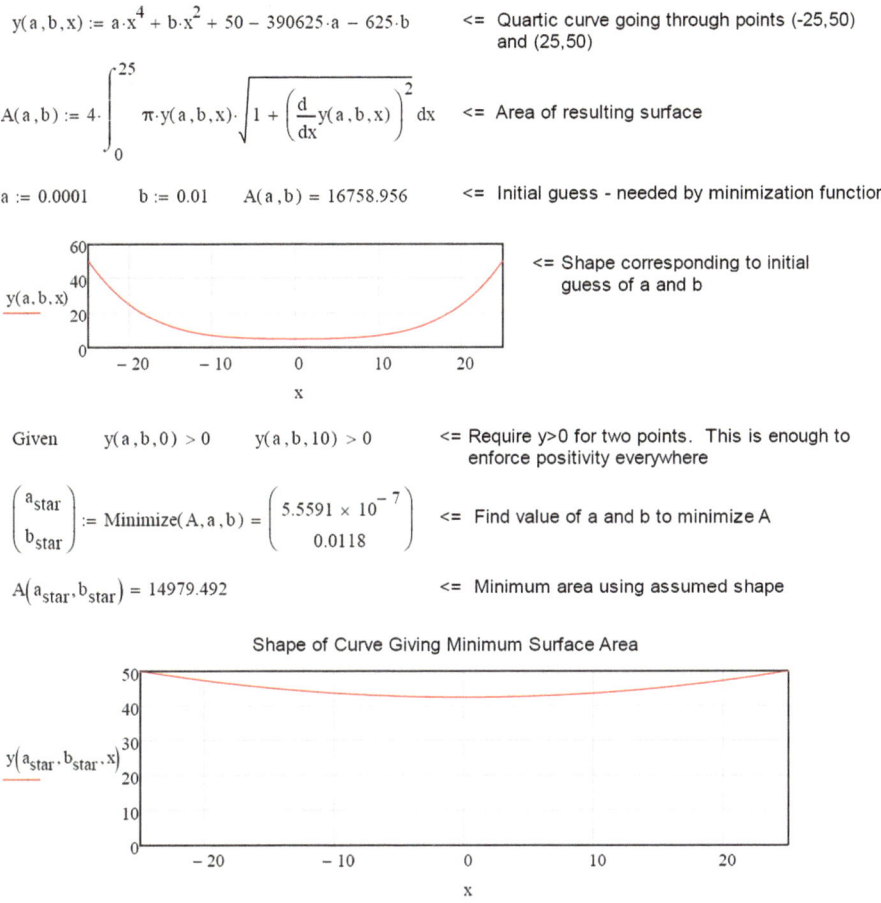

$$y(a,b,x) := a \cdot x^4 + b \cdot x^2 + 50 - 390625 \cdot a - 625 \cdot b \qquad \Leftarrow \text{ Quartic curve going through points (-25,50)}$$
and (25,50)

$$A(a,b) := 4 \cdot \int_0^{25} \pi \cdot y(a,b,x) \cdot \sqrt{1 + \left(\frac{d}{dx} y(a,b,x)\right)^2} \, dx \qquad \Leftarrow \text{ Area of resulting surface}$$

$a := 0.0001 \qquad b := 0.01 \qquad A(a,b) = 16758.956 \qquad \Leftarrow \text{ Initial guess - needed by minimization function}$

$\Leftarrow$ Shape corresponding to initial guess of a and b

Given $\quad y(a,b,0) > 0 \qquad y(a,b,10) > 0 \qquad \Leftarrow$ Require y>0 for two points. This is enough to enforce positivity everywhere

$$\begin{pmatrix} a_{star} \\ b_{star} \end{pmatrix} := \text{Minimize}(A,a,b) = \begin{pmatrix} 5.5591 \times 10^{-7} \\ 0.0118 \end{pmatrix} \qquad \Leftarrow \text{ Find value of a and b to minimize A}$$

$A(a_{star}, b_{star}) = 14979.492 \qquad\qquad \Leftarrow \text{ Minimum area using assumed shape}$

Shape of Curve Giving Minimum Surface Area

Fig. 4.10 Soap film problem with quartic assumed function

4.4   Solution Methods

The two-variable lifeguard example above showed the most direct solution method, that of setting $\nabla T = 0$ and solving the resulting set of equations. In practical optimization problems, it's unusual to have a closed form expression for the objective function. As a result, there are whole families of minimization algorithms that work on multivariable problems.

What follows is a very short selection of search methods for finding minimum values of multivariable objective functions. As stated earlier, the purpose of this book is to show how to recognize and set up optimization problems rather than to present a wide variety of solution methods. The methods presented here are simple and robust, but not very efficient. This is acceptable because they behave, from the user's point of view, similarly to the more efficient methods programmed into commercial software. We can start with the multivariable version of the Monte Carlo method.

4.4.1 Monte Carlo Method

As with single-variable problems, the simplest possible approach is to generate a set of random points in design space, calculate objective function values for each point, and just take the lowest one. The big idea behind the Monte Carlo method is that a small number of widely separated random points in design space might include a point that is acceptably close to the minimum. Figure 4.11 shows the Mathcad calculation using a uniform distribution for the random numbers. Even though this method uses a small number of random guesses, it found a point reasonably close to the optimum.

A MATLAB function that performs the same calculations follows:

```
function Monte_Carlo(N)

%  This function implements a simple Monte Carlo solution for
%  the two variable Lifeguard Problem.
%  N is the number of randomm points

clc  % Clear screen

% Define objective function using an anonymous function
T = @(x1,x2) 1/7*sqrt(50^2+x1.^2)+1/4*sqrt(20^2+(x2-x1).^2)+...
    1/2*sqrt(30^2+(100-x2).^2);

% Make contour plot
x1=0:100; x2=x1;
[X1,X2]=meshgrid(x1,x2);
Time=T(X1,X2);
contour(x1,x2,Time);grid;hold on
xlabel('x_1'); ylabel('x_2'); title(['N = ' num2str(N) ])

% Make lists of random numbers with uniform distribution
x1=rand(N,1)*100;
x2=rand(N,1)*100;
plot(x1,x2,'ob')

% Calculate corresponding times
time=T(x1,x2);

% Find minimum time and its vector address
[T_star,i]=min(time);

plot(x1(i),x2(i),'+r') % Put a small + at the minimum of random points
plot(81.326,92.472,'+b', 'MarkerSize',15) % Mark global minimum
hold off

disp(['T* = ' num2str(T_star) ])
disp(['x1* = ' num2str(x1(i)) ])
disp(['x2* = ' num2str(x2(i)) ])

return
```

Monte Carlo Solution for Two Variable Lifeguard Problem   ORIGIN≡ 1

$$T(x_1, x_2) := \frac{1}{7}\sqrt{50^2 + x_1^2} + \frac{1}{4}\sqrt{20^2 + (x_2 - x_1)^2} + \frac{1}{2}\sqrt{30^2 + (100 - x_2)^2}$$

N := 20   <= Select the number of random points

i := 1..N   <= Range variable going from 1 to N

$x1_i$:= rnd (100) $x2_i$:= rnd (100)   <= Make N random values of x1 and x2. Random
                                                         values are in the range 0-100.

Tmc_i := $T(x1_i, x2_i)$ <= Make the corresponding N random values of T

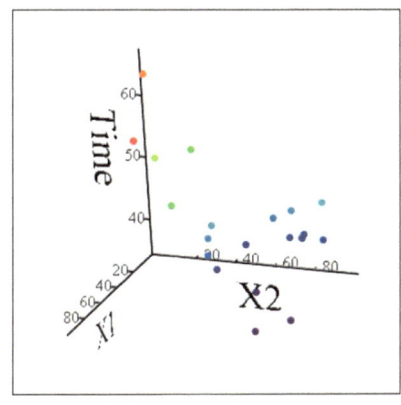

(x1, x2, Tmc)

A := augment(x1, x2)   A := augment(A, Tmc)   <= Matrix A has columns (x1,x2,T)

csort (A, 3)   <= Sort A according to 3rd column, T

		1	2	3
A =	1	0.127	87.597	45.806
	2	19.332	95.59	42.529
	3	58.501	53.934	...

<= Shortest time corresponding to
   random values of x1 and x2

$x1_{star}$:= $A_{1,1}$ = 0.127 $x2_{star}$:= $A_{1,2}$ = 87.597 $T(x1_{star}, x2_{star})$ = 45.806

Fig. 4.11 Monte Carlo solution to two-variable lifeguard problem

Figure 4.12 shows the surface plot for a solution with 20 random points using a
uniform distribution. The test output to the command window is

```
T*  = 35.0756
x1* = 72.9036
x2* = 87.3536
```

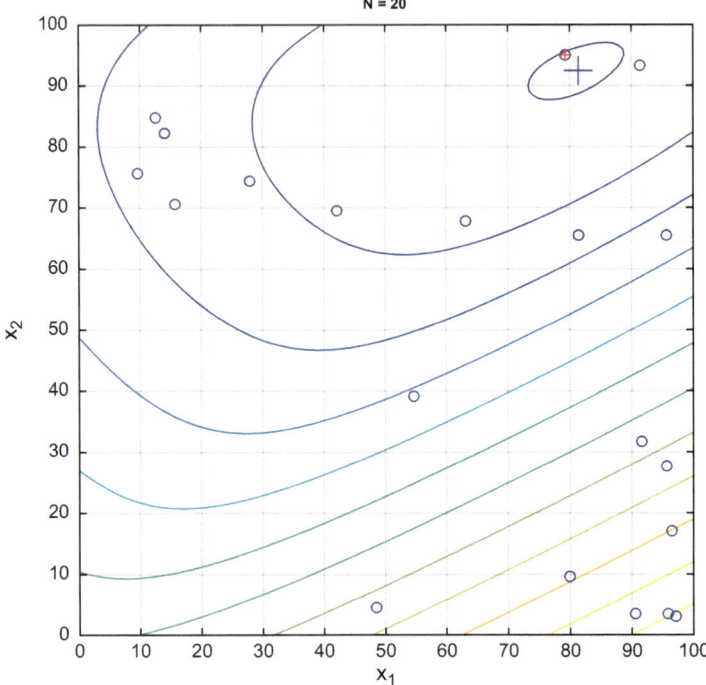

Fig. 4.12 Design space from the MATLAB Monte Carlo solution, $N = 20$

The estimated minimum time is less than 1% higher than the actual minimum. Clearly, using more points increases the chance of finding a point close to the minimum.

4.4.2   Marching Grid Algorithm

A key step in improving efficiency over the Monte Carlo method is to search design space in a deliberate way from a given starting point. An exceedingly simple and robust method is the marching grid algorithm. It is basically the multivariable analogy to the binary search algorithm [3].

The overarching idea is that a grid containing a small number of points can be moved in steps through design space. Each step is typically half the size of the grid. The objective function values are calculated for each point in the grid and every successive step centers the grid on the minimum value of the objective function. When the grid stops moving, that means the minimum lies within the grid. The size of the grid is then reduced and the process repeated until some exit criteria is satisfied.

Start by identifying a starting point, (x_0, y_0), and step sizes Δx and Δy. For simplicity, we will assume $\Delta x = \Delta y$. Assign a 3×3 grid of points as shown in Fig. 4.13, and find the minimum objective function value of those nine points. For this example, $\Delta x = \Delta y = 20$ m and the starting point is $x_0 = y_0 = 50$ m.

The location of the minimum objective function value for the first grid becomes the center of the second grid and so on. When the grid stops moving, that means the minimum objective function value is near the center of the grid, and it is time to reduce the step size. Figure 4.14 shows the second grid calculation with the center at $x = x_0$ and $y = y_0 + \Delta y$. This is point (50, 70). In optimization terminology, our current estimate of the minimum is $x^* = 50$ m, $y^* = 70$ m.

Marching Grid Example

$$T(x1, x2) := \frac{1}{7} \cdot \sqrt{50^2 + x1^2} + \frac{1}{4} \cdot \sqrt{20^2 + (x2 - x1)^2} + \frac{1}{2} \cdot \sqrt{30^2 + (100 - x2)^2}$$

Select Starting Point and Step Size

$x_0 := 50$ $\qquad$ $y_0 := 50$ $\qquad$ $\Delta x := 20$ $\qquad$ $\Delta y := 20$

Calculate Points for Initial Grid

$T(x_0 - \Delta x, y_0 + \Delta y) = 40.723$ $\qquad$ $T(x_0, y_0 + \Delta y) = 38.386$ $\qquad$ $T(x_0 + \Delta x, y_0 + \Delta y) = 38.502$

$T(x_0 - \Delta x, y_0) = 44.556$ $\qquad$ $T(x_0, y_0) = 44.256$ $\qquad$ $T(x_0 + \Delta x, y_0) = 48.515$

$T(x_0 - \Delta x, y_0 - \Delta y) = 51.409$ $\qquad$ $T(x_0, y_0 - \Delta y) = 55.251$ $\qquad$ $T(x_0 + \Delta x, y_0 - \Delta y) = 61.548$

$x_{star} := x_0 = 50$ $\qquad$ $y_{star} := y_0 + \Delta y = 70$ $\qquad$ <= Assign estimate for x* and y*

Fig. 4.13 First step in marching grid algorithm

Marching Grid Example - Iteration 2

$x_1 := x_{star}$ $\qquad$ $y_1 := y_{star}$

$T(x_1 - \Delta x, y_1 + \Delta y) = 39.953$ $\qquad$ $T(x_1, y_1 + \Delta y) = 37.093$ $\qquad$ $T(x_1 + \Delta x, y_1 + \Delta y) = 35.171$

$T(x_1 - \Delta x, y_1) = 40.723$ $\qquad$ $T(x_1, y_1) = 38.386$ $\qquad$ $T(x_1 + \Delta x, y_1) = 38.502$

$T(x_1 - \Delta x, y_1 - \Delta y) = 44.556$ $\qquad$ $T(x_1, y_1 - \Delta y) = 44.256$ $\qquad$ $T(x_1 + \Delta x, y_1 - \Delta y) = 48.515$

$x_{star} := x_1 + \Delta x = 70$ $\qquad$ $y_{star} := y_1 + \Delta y = 90$ $\qquad$ <= Assign estimate for x* and y*

Fig. 4.14 Second step in marching grid

Marching Grid Example - Iteration 3

$$x_2 := x_{star} \qquad y_2 := y_{star}$$

$T(x_2 - \Delta x, y_2 + \Delta y) = 41.724$ $\qquad$ $T(x_2, y_2 + \Delta y) = 39.281$ $\qquad$ $T(x_2 + \Delta x, y_2 + \Delta y) = 37.59$

$T(x_2 - \Delta x, y_2) = 37.093$ $\qquad$ $T(x_2, y_2) = 35.171$ $\qquad$ $T(x_2 + \Delta x, y_2) = 35.519$

$T(x_2 - \Delta x, y_2 - \Delta y) = 38.386$ $\qquad$ $T(x_2, y_2 - \Delta y) = 38.502$ $\qquad$ $T(x_2 + \Delta x, y_2 - \Delta y) = 42.992$

$$x_{star} := x_2 = 70 \qquad\qquad y_{star} := y_2 = 90 \qquad\qquad \Leftarrow \text{Assign estimate for } x^* \text{ and } y^*$$

Fig. 4.15 Third step in marching grid

Fig. 4.16 Graphical representation of marching grid

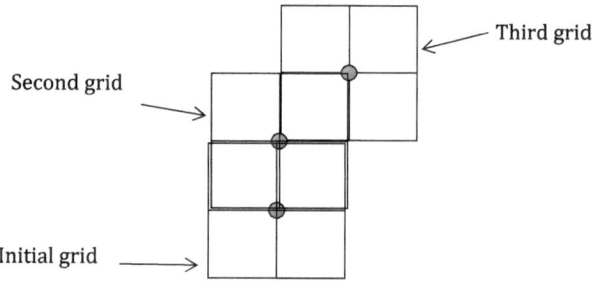

Second grid

Third grid

Initial grid

The new estimate of the minimum is point (70, 90) and will be the center of the next grid. It will take one more iteration to show that the grid has stopped moving as shown in Fig. 4.15.

Marching grid is perhaps the easiest search method to understand graphically. Figure 4.16 shows how the calculation grid has moved for the first three iterations. There is a slight offset in the grid locations, so you can identify them more easily. Also, the circle is the center location of each grid.

Since the minimum value of the objective function lies at the center of the grid, it has stopped moving. This means that the minimum objective function value likely lies within the grid, and the step size needs to be reduced. We will cut the step sizes in half and continue again until the grid stops moving, as shown in Fig. 4.17.

After the fifth iteration, the estimate of the minimum is $x^* = 80$ and $y^* = 90$, and the grid has stopped moving. The next step would be to cut the step size in half again and continue the iteration. Iteration stops when the grid size goes below a predetermined limit. The current estimate of T^* is about 0.1% more than the actual minimum, so it's fine to stop here.

Marching grid is very robust in the sense that it will almost always find at least a local minimum. However, it is not always very efficient. If the grid size is not chosen very carefully, it could take a large number of objective function evaluations to find

Marching Grid Example - Iteration 4

$$x_3 := x_{star} \qquad y_3 := y_{star} \qquad \Delta x := \frac{\Delta x}{2} = 10 \qquad \Delta y := \frac{\Delta y}{2} = 10$$

$$T(x_3 - \Delta x, y_3 + \Delta y) = 37.338 \qquad T(x_3, y_3 + \Delta y) = 36.303 \qquad T(x_3 + \Delta x, y_3 + \Delta y) = 35.548$$

$$T(x_3 - \Delta x, y_3) = 35.983 \qquad T(x_3, y_3) = 35.171 \qquad T(x_3 + \Delta x, y_3) = 34.879$$

$$T(x_3 - \Delta x, y_3 - \Delta y) = 36.256 \qquad T(x_3, y_3 - \Delta y) = 35.907 \qquad T(x_3 + \Delta x, y_3 - \Delta y) = 36.505$$

$$x_{star} := x_3 + \Delta x = 80 \qquad y_{star} := y_3 = 90 \qquad \text{<= Assign estimate for x* and y*}$$

Marching Grid Example - Iteration 5

$$x_4 := x_{star} \qquad y_4 := y_{star}$$

$$T(x_4 - \Delta x, y_4 + \Delta y) = 36.303 \qquad T(x_4, y_4 + \Delta y) = 35.548 \qquad T(x_4 + \Delta x, y_4 + \Delta y) = 35.298$$

$$T(x_4 - \Delta x, y_4) = 35.171 \qquad T(x_4, y_4) = 34.879 \qquad T(x_4 + \Delta x, y_4) = 35.519$$

$$T(x_4 - \Delta x, y_4 - \Delta y) = 35.907 \qquad T(x_4, y_4 - \Delta y) = 36.505 \qquad T(x_4 + \Delta x, y_4 - \Delta y) = 38.326$$

$$x_{star} := x_4 = 80 \qquad y_{star} := y_4 = 90 \qquad \text{<= Assign estimate for x* and y*}$$

Fig. 4.17 Fourth and fifth steps in marching grid

the minimum. An even more serious limitation is that the method quickly becomes unworkable as the number of design variables increases. The number of points in the grid is 3^n, where n is the number of design variables. An unacceptably large number of objective function evaluations is needed for all but the smallest problems.

Figure 4.18 shows a pseudo flow chart for the marching grid algorithm. Note that, for simplicity, this description assumes Δx and Δy will change sizes at the same time. There are many possible variations on this basic method, and one of these would be to let Δx and Δy vary separately.

The advantages of the marching grid algorithm are that it is very robust and it does not require calculating gradients. The disadvantages are that it converges slowly and that it becomes unwieldy for more than a few design variables. The nine-point grid for a two-variable problem becomes a 27-point cube for a three-variable problem and an 81-point hypercube for a four-variable problem. Clearly something less cumbersome and more efficient is needed.

4.4.3   Steepest Descent Method

Steepest descent is a method that uses information from the gradient (the equivalent of the derivative for more than one variable) of the objective function to find a search

Select x_0, y_0, Δx, Δy, Δx_{min}, Δy_{min}

Calculate first grid

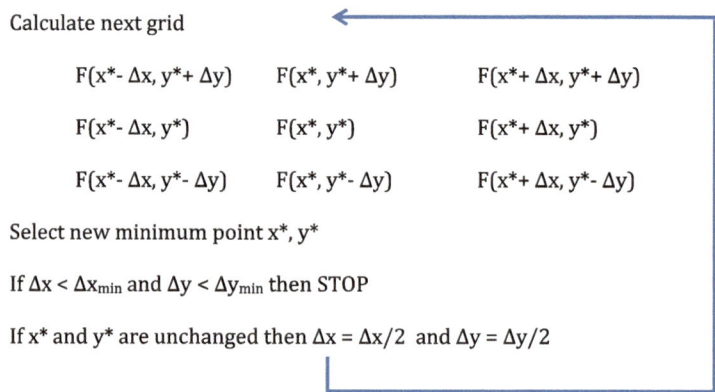

Select x^* and y^* based on minimum value of F

If $x^* = x_0$ and $y^* = y_0$ then $\Delta x = \Delta x/2$ and $\Delta y = \Delta y/2$

Select new minimum point x^*, y^*

If $\Delta x < \Delta x_{min}$ and $\Delta y < \Delta y_{min}$ then STOP

If x^* and y^* are unchanged then $\Delta x = \Delta x/2$ and $\Delta y = \Delta y/2$

Fig. 4.18 Pseudo flow chart for marching grid

direction through design space. Steepest descent is robust in that it will always find at least a local minimum and will not oscillate between two points in design space as some other methods can. Also, it is more efficient than marching grid in that it will generally find a minimum with fewer objective function evaluations. There are more efficient methods than the steepest descent, though sometimes they are less robust.

The method slides downhill from some starting point with the downhill direction being defined by the gradient, usually denoted as ∇f. It is iterative, so the downhill direction is calculated by gradients evaluated for each iteration [4].

The gradient is a vector that encodes two very important pieces of information. The direction of the vector is the direction of maximum slope, and the magnitude of the vector is the value of the slope in that direction. Remember that the slope of a surface depends not only on the position but also on the direction one is looking.

Imagine yourself standing on a hillside as idealized in Fig. 4.19. Now imagine yourself turning to face uphill. The direction you are facing corresponds to the direction of the gradient at that point. The slope in the direction you are facing is the magnitude of the gradient at that point. Now imagine yourself turning 90° to the left or right. The slope is now zero, even though your location hasn't changed. Finally, imagine yourself turning another 90° to face downhill. That direction is the

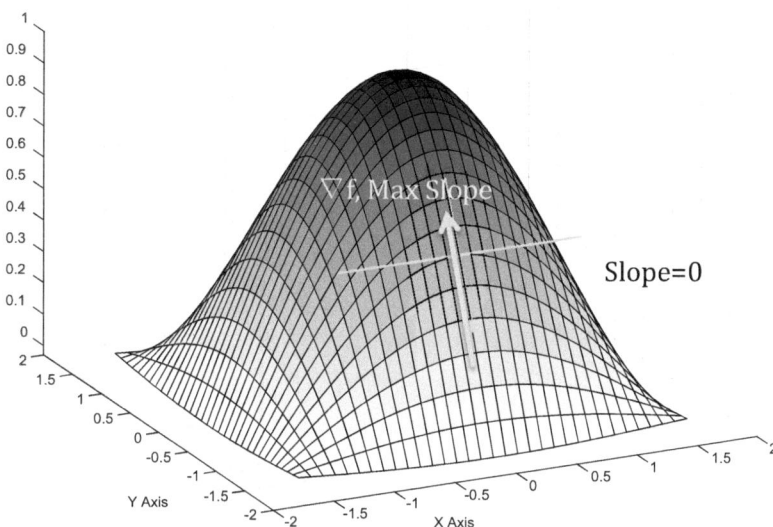

Fig. 4.19 A mathematical "hill" showing maximum and zero slope at one point

negative of the gradient vector, and the slope is the negative of the magnitude of the gradient.

For two-variable problems, the gradient is apparent on a contour plot. The lines on a contour plot connect points of equal height, so the slope along a contour line is zero. Thus, the gradient at any point is perpendicular to the contour line at that point. Note that direction and magnitude of the gradient depend on where it is calculated. Figure 4.20 shows an example. The arrows only indicate direction of the negative of the gradient. They are not scaled to show magnitude.

The idea behind the steepest descent method is simple. It selects a search direction in design space and finds a minimum value of the objective function in that direction – the point at which the search direction is parallel to the contour lines. It then selects a new direction and repeats the process, stopping when the exit criteria have been satisfied.

To start, the user selects an initial point and calculates the gradient at that point. For the two-variable lifeguard problem, the gradient is

$$\nabla T = \left\{ \begin{array}{c} \dfrac{\partial T}{\partial x_1} \\[2mm] \dfrac{\partial T}{\partial x_2} \end{array} \right\} \tag{4.13}$$

The negative of the gradient (the direction of steepest descent) becomes the first search direction, S^1.

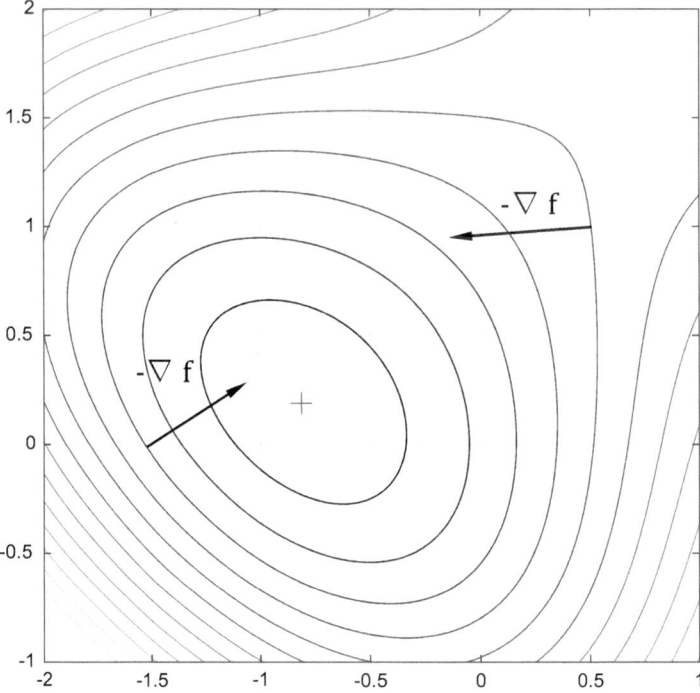

Fig. 4.20 Gradient on a contour plot

$$S^1 = -\nabla T = -\left\{ \begin{array}{c} \dfrac{\partial T}{\partial x_1} \\[2mm] \dfrac{\partial T}{\partial x_2} \end{array} \right\} = \left\{ \begin{array}{c} s_1^1 \\[2mm] s_2^1 \end{array} \right\} \tag{4.14}$$

Note the use of the superscript to identify iteration number and the subscript to identify the variable number. This nomenclature lets us work with an arbitrary number of design variables and an arbitrary number of iterations. Just remember that the superscript is not an exponent. Fortunately, there will not be a need to raise a search vector to a power.

Once a search direction through design space has been chosen, a 1D search is done in that direction to find a minimum value of the objective function. Once the minimum is identified, another gradient is calculated and so on. It is important to realize that no matter how many design variables the problem has, selecting a search direction reduces the problem to a single variable that describes the distance along the search direction. We'll call that distance d. For our two-variable problem, the relationship between the design variables and d for the first iteration is

Steepest Descent Example

$$T(x1, x2) := \frac{1}{7} \cdot \sqrt{50^2 + x1^2} + \frac{1}{4} \cdot \sqrt{20^2 + (x2 - x1)^2} + \frac{1}{2} \cdot \sqrt{30^2 + (100 - x2)^2}$$

Select Starting Point $\quad\quad x1_0 := 50 \quad\quad x2_0 := 50 \quad\quad T\left(x1_0, x2_0\right) = 44.256$

Find the first search direction

$$S(x1, x2) := -\begin{pmatrix} \dfrac{d}{dx1} T(x1, x2) \\[2mm] \dfrac{d}{dx2} T(x1, x2) \end{pmatrix} \rightarrow \begin{bmatrix} -\dfrac{x1}{7 \cdot \sqrt{x1^2 + 2500}} - \dfrac{2 \cdot x1 - 2 \cdot x2}{8 \cdot \sqrt{(x1 - x2)^2 + 400}} \\[4mm] \dfrac{2 \cdot x1 - 2 \cdot x2}{8 \cdot \sqrt{(x1 - x2)^2 + 400}} - \dfrac{2 \cdot x2 - 200}{4 \cdot \sqrt{(x2 - 100)^2 + 900}} \end{bmatrix}$$

$$S1 := S\left(x1_0, x2_0\right) = \begin{pmatrix} -0.101 \\ 0.429 \end{pmatrix} \quad\quad \text{<= First search direction}$$

Set up the 1-D search problem

$$x1d(d) := x1_0 + d \cdot S1_0 \quad\quad x2d(d) := x2_0 + d \cdot S1_1 \quad\quad Td(d) := T(x1d(d), x2d(d))$$

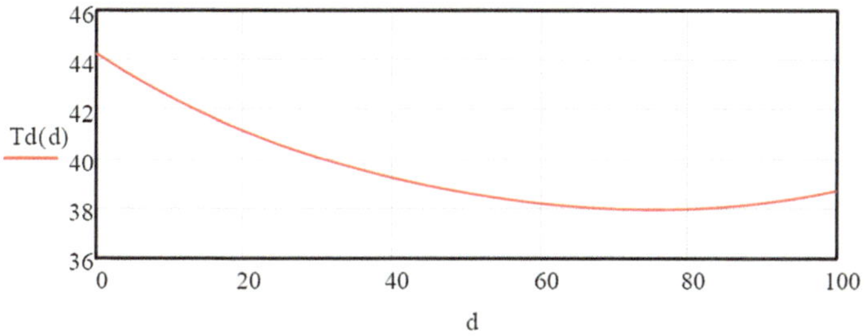

Now find the minimum along the search direction

dd := 0 $\quad\quad$ <= Initial guess

$d_{star} :=$ Minimize(Td, dd) = 75.275 $\quad\quad$ <= Minimum in the search direction

$x1_{star} := x1d\left(d_{star}\right) = 42.396 \quad\quad x2_{star} := x2d\left(d_{star}\right) = 82.274 \quad\quad T_{star} := Td\left(d_{star}\right) = 37.941$

Fig. 4.21 First iteration of the steepest descent method

$$\begin{aligned} x_1 &= x_0 + dS_1^l \\ y_1 &= y_0 + dS_2^l \end{aligned} \quad\quad (4.15)$$

Figure 4.21 shows the first iteration of the steepest descent using the starting point $x_0 = y_0 = 50$. The starting point is essentially arbitrary, though it is often conditioned by some knowledge of the nature of the problem. In this case, it seems obvious that the minimum answers must lie in the range [0,100], so picking a middle point is reasonable.

$$S2 := S\left(x1_{star}, x2_{star}\right) = \begin{pmatrix} 0.131 \\ 0.031 \end{pmatrix} \qquad \text{<= Second search direction}$$

$$x1d(d) := x1_{star} + d \cdot S2_0 \qquad x2d(d) := x2_{star} + d \cdot S2_1 \qquad Td(d) := T(x1d(d), x2d(d))$$

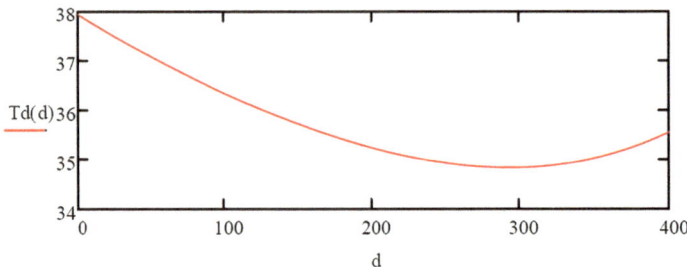

Now find the minimum along the search direction

$$d_{star} := \text{Minimize}(Td, dd) = 293.259 \qquad \text{<= Minimum in the search direction}$$

Define second estimate for x1*, x2* and T*

$$x1_{star} := x1d\left(d_{star}\right) = 80.837 \qquad x2_{star} := x2d\left(d_{star}\right) = 91.331 \qquad T_{star} := Td\left(d_{star}\right) = 34.839$$

Fig. 4.22 Second iteration of steepest descent

In one iteration, the steepest descent has moved from an arbitrary point in design space to a point that gives an objective function value within 10% of the minimum. The second iteration starts at our current estimate of $x*$ and $y*$ and calculates a new search direction as shown in Fig. 4.22.

Note that d does not correspond directly to a distance across the sand or water. Rather, it is a scaled distance through design space.

Figure 4.23 shows the path of the first two iterations through design space. It is clear that the first and second search directions are perpendicular to one another. Note also how close we are now to the minimum value of the objective function.

Practically speaking, we could stop now, but let's do one more iteration. Figure 4.24 shows the third iteration.

The algorithm has essentially converged on the minimum value of the objective function and more iterations will be of little use. One way to show this graphically is a convergence plot as shown in Fig. 4.25. The horizontal axis is the iteration number, and the vertical axis is the objective function value.

One important feature of the steepest descent is that successive search directions are perpendicular to one another. This is easy to check using the definition of a vector dot product.

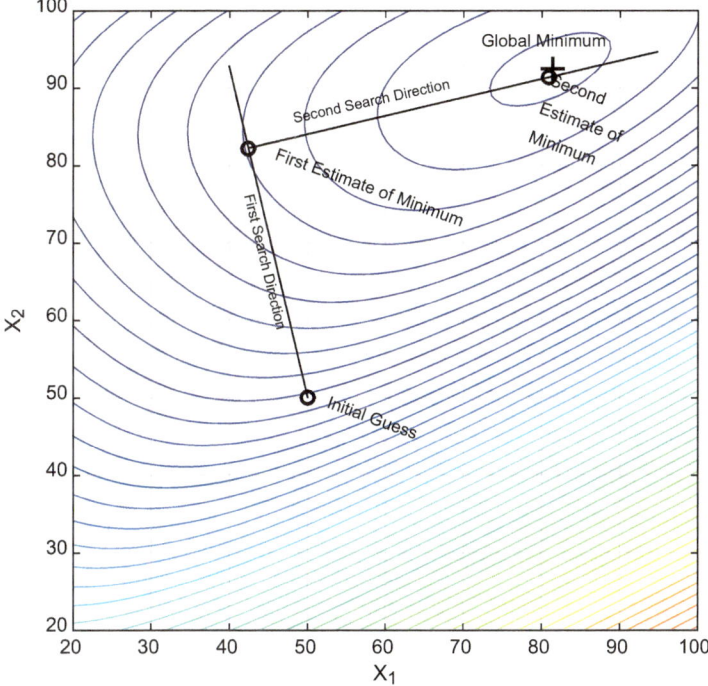

Fig. 4.23 Path of steepest descent through design space

$$A \cdot B = |A||B| \cos \theta \qquad (4.16)$$

For the two-variable lifeguard example, the dot product of the first and second search directions, $S^1 \cdot S^2$, is

$$S^1 \cdot S^2 = -0.101 \times 0.131 + 0.429 \times 0.031 = 0 \qquad (4.17)$$

In graphical terms, the steepest descent search direction is perpendicular to the curves of constant height on the contour plot like the one in Fig. 4.2. The minimum value, d^*, in the 1D search occurs when the search direction is parallel to a curve of constant height. Since the next search direction is perpendicular to the contour curve, it must be perpendicular to the previous search direction. All three of the steepest descent search paths are shown in Fig. 4.26. If the search for the minimum were to be continued, the path to the global minimum of the objective function would continue in zigzag fashion.

In this case, the steepest descent method found the minimum in a small number of iterations. However, there are cases in which this method is forced to traverse a diagonal that is not aligned with a search direction. When this happens, the method is forced to zigzag toward the minimum. While it certainly works, this approach is not

Third Iteration

$$S3 := S\left(x1_{star}, x2_{star}\right) = \begin{pmatrix} -5.337 \times 10^{-3} \\ 0.023 \end{pmatrix} \qquad \text{<= Second search direction}$$

$x1d(d) := x1_{star} + d \cdot S3_0 \qquad x2d(d) := x2_{star} + d \cdot S3_1 \qquad Td(d) := T(x1d(d), x2d(d))$

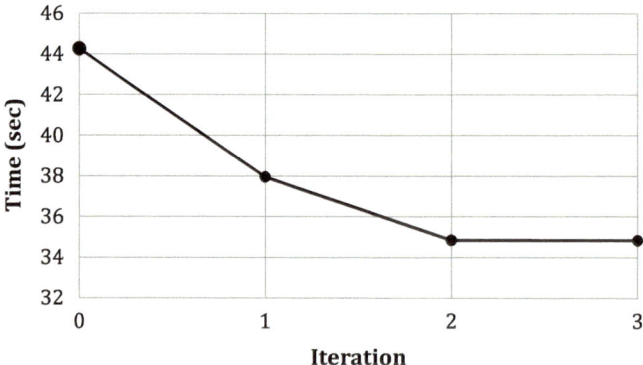

Now find the minimum along the search direction

$d_{star} := \text{Minimize}(Td, dd) = 38.005 \qquad \text{<= Minimum in the search direction}$

Define third estimate for x1*, x2* and T*

$x1_{star} := x1d\left(d_{star}\right) = 80.634 \qquad x2_{star} := x2d\left(d_{star}\right) = 92.192 \qquad T_{star} := Td\left(d_{star}\right) = 34.829$

Fig. 4.24 Third iteration of the steepest descent

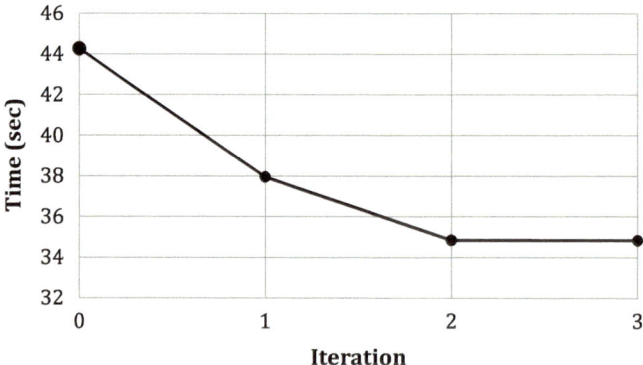

Fig. 4.25 Convergence plot for the steepest descent

very efficient. There is an example in the next section, where a more efficient method is introduced.

Finally, Fig. 4.27 shows the pseudo flow chart for the steepest descent algorithm. Note that every iteration of the steepest descent requires a 1D search of the type discussed earlier. Also, it often makes sense to express Δx_{min} and ΔF_{min} in percentages.

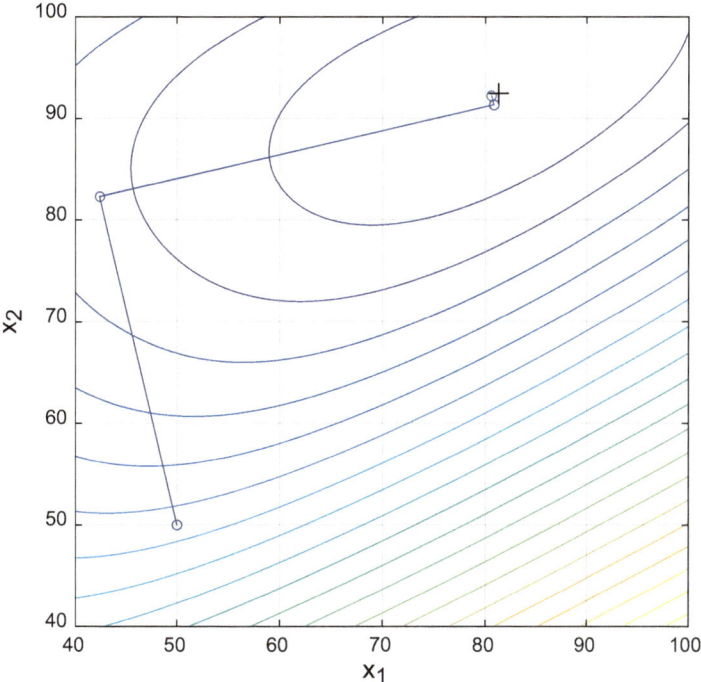

Fig. 4.26 Search path for the steepest descent

Select $x_1^0, x_2^0, \cdots, x_n^0, \Delta x_{min}, \Delta F_{min}$

m=0

$S^{m+1} = -\nabla F(x_1^m, x_2^m, \cdots, x_n^m)$

$xd_1 = x_1^m + d \cdot S_1^{m+1}, xd_2 = x_2^m + d \cdot S_2^{m+1}, \cdots, xd_n = x_n^m + d \cdot S_n^{m+1}$

$F_d(d) = F(xd_1, xd_2, \cdots, xd_n)$

1-D search to find d* and F_d*

$x_1^* = x_1^m + d^* \cdot S_1^{m+1}, x_2^* = x_2^m + d^* \cdot S_2^{m+1}, \cdots, x_n^* = x_n^m + d^* \cdot S_n^{m+1}$

$F^* = F(x_1^*, x_2^*, \cdots, x_n^*)$

If change in $x_1, x_2, \ldots, x_n < \Delta x_{min}$ and change in F < F_{min} then STOP

m=m+1

$x_1^m = x_1^*, x_2^m = x_2^*, \cdots$

Fig. 4.27 Pseudo flow chart for the steepest descent

4.4.4 Conjugate Gradient Method

In the steepest descent method, successive search directions are perpendicular. Thus, after finding the first search direction, we know all the following ones, except for the sign. However, there is no reason to think that the most efficient succession of search directions should all be perpendicular. The problem, of course, is that a better choice of search direction likely requires more information than the gradient at one point.

Since previous search directions have already been calculated, it might be possible to use information about previous search directions to find the next one. Many ways have been proposed, and one of the most popular is the conjugate gradient method as proposed by Fletcher and Reeves [5]. In this method, the nth search direction is defined as

$$S^n = -\nabla F\left(x^{n-1}\right) + \frac{\left|\nabla F(x^{n-1})\right|^2}{\left|\nabla F(x^{n-2})\right|^2} S^{n-1} \tag{4.18}$$

The first search direction is the same as in steepest descent since there is no previous search direction on which to build. However, the second search direction and all those following will be different.

The more general form of the conjugate gradient method defines the nth search direction as

$$S^n = -\nabla F\left(x^{n-1}\right) + \beta_n S^{n-1} \tag{4.19}$$

There are several definitions of β_n in the literature, with the Fletcher-Reeves definition being just one example.

The conjugate gradient method tends to be superior to the steepest descent in situations where the steepest descent is forced to zigzag through design space. In cases where this is not necessary, conjugate gradient may offer little advantage over steepest descent.

Because the lifeguard problem has been a standard problem, let's apply the conjugate gradient to it first. Specifically, let's look at the performance of the conjugate gradient method in solving the two-variable lifeguard problem. Figure 4.28 shows the first two iterations of the conjugate gradient method using the same starting point, $x_0 = 50$ and $y_0 = 50$, as before.

Note that the second estimate of the minimum value of the objective function, T^*, is actually higher for conjugate gradient than it was for the steepest descent: 35.972 for conjugate gradient vs. 34.839 for steepest descent. To continue, let's continue to the third iteration, shown in Fig. 4.29.

Again, conjugate gradient converges no faster than steepest descent. After the third iteration, the steepest descent estimates $T^* = 34.829$, while conjugate gradient estimates $T^* = 34.902$.

Finally, Fig. 4.30 shows the paths taken by the steepest descent and conjugate gradient methods. While conjugate gradient takes a different path, it isn't, in this case, a better one.

Two Variable Lifeguard Problem - Fletcher-Reeves Conjugate Gradient Method

$$T(x1,x2) := \frac{1}{7}\cdot\sqrt{50^2 + x1^2} + \frac{1}{4}\cdot\sqrt{20^2 + (x2 - x1)^2} + \frac{1}{2}\cdot\sqrt{30^2 + (100 - x2)^2}$$

Iteration 1 is identical to steepest descent

$$\text{gradT}(x1,x2) := \begin{pmatrix} \dfrac{d}{dx1}T(x1,x2) \\[2ex] \dfrac{d}{dx2}T(x1,x2) \end{pmatrix} \rightarrow \begin{bmatrix} \dfrac{x1}{7\cdot\sqrt{x1^2 + 2500}} + \dfrac{2\cdot x1 - 2\cdot x2}{8\cdot\sqrt{(x1 - x2)^2 + 400}} \\[3ex] \dfrac{2\cdot x2 - 200}{4\cdot\sqrt{(x2 - 100)^2 + 900}} - \dfrac{2\cdot x1 - 2\cdot x2}{8\cdot\sqrt{(x1 - x2)^2 + 400}} \end{bmatrix}$$

$x1_0 := 50 \qquad x2_0 := 50 \qquad S1 := -\text{gradT}\left(x1_0, x2_0\right) = \begin{pmatrix} -0.101 \\ 0.429 \end{pmatrix} \qquad T\left(x1_0, x2_0\right) = 44.256$

$x1d(d) := x1_0 + d\cdot S1_0 \qquad x2d(d) := x2_0 + d\cdot S1_1 \qquad Td(d) := T(x1d(d), x2d(d))$

$dd := 0 \qquad\qquad d_{star} := \text{Minimize}(Td, dd) = 75.275$

$x1_{star} := x1d\left(d_{star}\right) = 42.396 \qquad x2_{star} := x2d\left(d_{star}\right) = 82.274 \qquad T_{star} := Td\left(d_{star}\right) = 37.941$

First estimate of the minimum => $\qquad x1_1 := x1_{star} = 42.396 \qquad x2_1 := x2_{star} = 82.274$

Second iteration uses modified search direction - conjugate gradient

$$S2 := -\text{gradT}\left(x1_1, x2_1\right) + \frac{\left(\left|\text{gradT}\left(x1_1, x2_1\right)\right|\right)^2}{\left(\left|\text{gradT}\left(x1_0, x2_0\right)\right|\right)^2}\cdot S1 = \begin{pmatrix} 0.122 \\ 0.071 \end{pmatrix}$$

$x1d(d) := x1_1 + d\cdot S2_0 \qquad x2d(d) := x2_1 + d\cdot S2_1 \qquad Td(d) := T(x1d(d), x2d(d))$

$d_{star} := \text{Minimize}(Td, dd) = 207.822 \qquad$ <= Minimum in the search direction

$x1_{star} := x1d\left(d_{star}\right) = 67.675 \qquad x2_{star} := x2d\left(d_{star}\right) = 97.02 \qquad T_{star} := Td\left(d_{star}\right) = 35.972$

Second estimate of the minimum $\qquad x1_2 := x1_{star} = 67.675 \qquad x2_2 := x2_{star} = 97.02$

Fig. 4.28 First two iterations of conjugate gradient method

The two-variable lifeguard problem is not challenging for the method of steepest descent, and the conjugate gradient method is not superior to it, at least for this starting point. Rather than just shrugging and moving on, let's try a problem for which the steepest descent method struggles and the conjugate gradient is more useful.

The Rosenbrock banana function (see Appendix B) is often used to demonstrate optimization algorithms since it is compact and there is often not a direct path from a starting point to the minimum. For this objective function, one would expect the

Third iteration uses modified search direction - conjugate gradient

$$S3 := -\mathrm{gradT}\left(x1_2, x2_2\right) + \frac{\left(\left|\mathrm{gradT}\left(x1_2, x2_2\right)\right|\right)^2}{\left(\left|\mathrm{gradT}\left(x1_1, x2_1\right)\right|\right)^2} \cdot S2 = \begin{pmatrix} 0.314 \\ -0.028 \end{pmatrix}$$

$\mathrm{x1d(d)} := x1_2 + d \cdot S3_0$ $\qquad$ $\mathrm{x2d(d)} := x2_2 + d \cdot S3_1$ $\qquad$ $\mathrm{Td(d)} := \mathrm{T}(\mathrm{x1d(d)}, \mathrm{x2d(d)})$

$\mathrm{dd} := 50$ $\qquad$ $\mathrm{d_{star}} := \mathrm{Minimize}(\mathrm{Td}, \mathrm{dd}) = 54.24$ $\qquad$ <= Minimum in the search direction

$\mathrm{x1_{star}} := \mathrm{x1d}\left(\mathrm{d_{star}}\right) = 84.693$ $\qquad$ $\mathrm{x2_{star}} := \mathrm{x2d}\left(\mathrm{d_{star}}\right) = 95.522$ $\qquad$ $\mathrm{T_{star}} := \mathrm{Td}\left(\mathrm{d_{star}}\right) = 34.902$

Third estimate of the minimum $\qquad$ $x1_3 := \mathrm{x1_{star}}$ $\qquad$ $x2_3 := \mathrm{x2_{star}}$

Fig. 4.29 Third iteration of conjugate gradient method

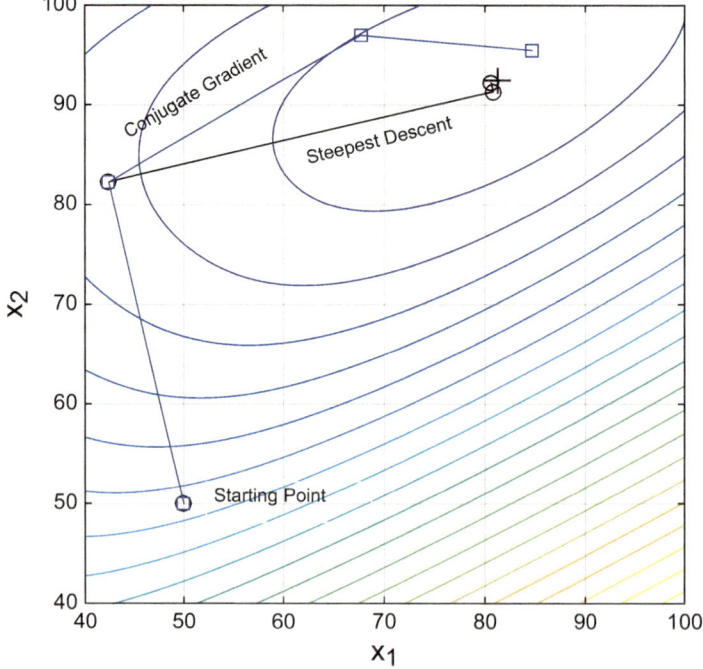

Fig. 4.30 Paths taken by conjugate gradient and steepest descent

steepest descent method to zigzag toward the minimum. Conversely, the conjugate gradient method might find a more direct route.

Figure 4.31 shows the path of both the steepest descent method and the conjugate gradient method from the starting point [5, 5] and progressing through the first six

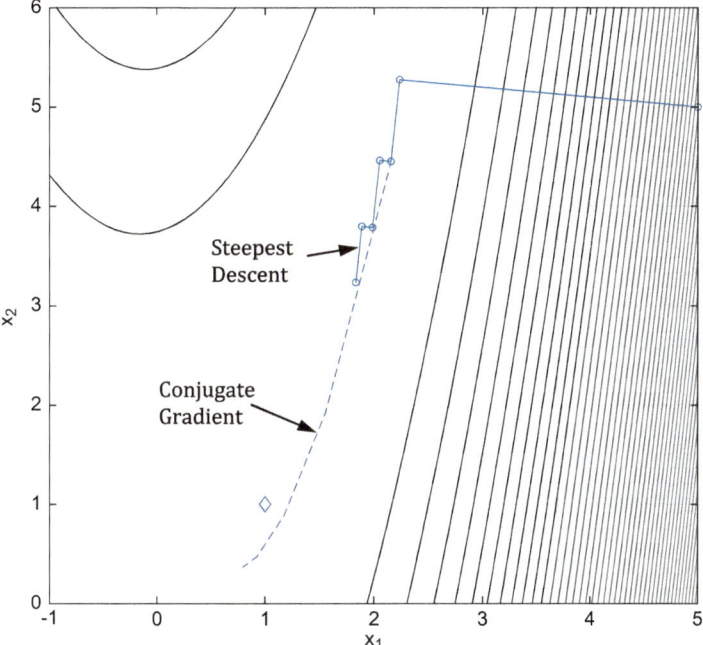

Fig. 4.31 Steepest descent and conjugate gradient methods

iterations. As always, the two are identical for the first iteration. In this case, they take very close to the same path for the second iteration as well. However, the third iteration shows the inherent limitation of the steepest descent method – in order to move along a diagonal, it must zigzag in perpendicular steps. The conjugate gradient method, in contrast, is able to move quickly toward the minimum point, at [1, 1] marked by the diamond.

The problem with the conjugate gradient method is clear in that it passes the minimum and will eventually have to turn around in order to reach it. It is well-known that the conjugate gradient method often performs better if, periodically, the search direction is calculated using the steepest descent.

Figure 4.32 shows the improvement when the fourth search direction in the conjugate gradient method is replaced by one calculated using the steepest descent. There is an abrupt shift to the left, after which conjugate gradient finds the minimum quickly. This type of hybrid method is not uncommon among practical implementations of minimization methods.

4.4.5 Newton's Method

The steepest descent and conjugate gradient methods work and are easy to implement, in part, because they rely only on first derivatives. That means that they work

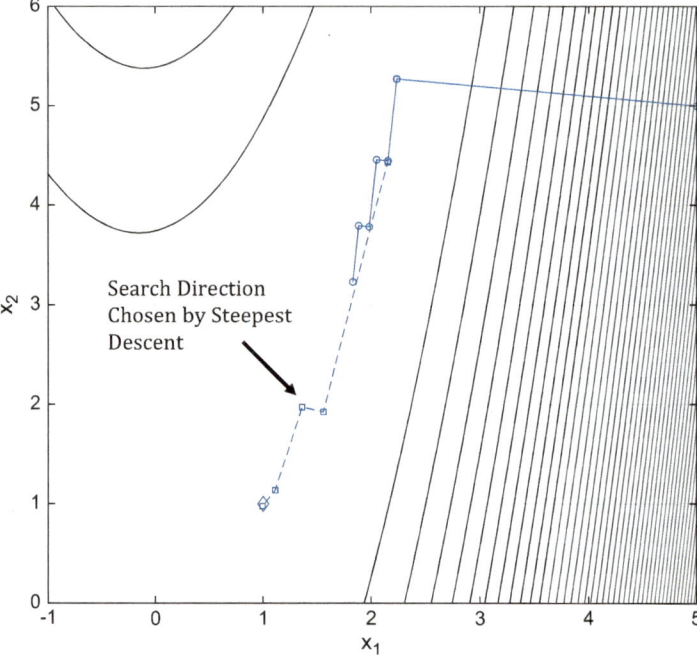

Fig. 4.32 Hybrid method compared to the steepest descent

using slopes, but not curvatures. It's easy to see that approximating a function with
both slopes and curvatures should work better than trying to approximate it using
only slopes. The most basic optimization method that uses both slopes and curvatures is called Newton's method [6]. The algorithm for Newton's method is identical
to the steepest descent except for the way the search direction is calculated.

To see how it works, let's start with a function of a single variable and extend the
approach to multiple variables. The process of using a slope to make a search
direction is mathematically equivalent to approximating the function with a
one-term Taylor series, so including curvature is equivalent to using a two-term
series. Equation (4.20) shows a two-term Taylor series approximation to a function
of one variable

$$f(x) = f(a + \Delta x) \approx f(a) + f'(a)\Delta x + \frac{1}{2}f''(a)\Delta x^2 \qquad (4.20)$$

where a is the point about which the function, $f(x)$, is expanded and $\Delta x = (x - a)$. If
the goal is to find the minimum or maximum value of $f(x)$, the approach every
calculus student learns is to set $df/dx = 0$ and solve for x. In this case, it is simpler to
take the derivative with respect to Δx.

$$f'(a) + f''(a)\Delta x = 0 \qquad (4.21)$$

Note that $\Delta x = x - a$ and a is a constant. Taking the derivative with respect to Δx is equivalent to taking the derivative with respect to x. Using Eq. (4.21), the estimate of the maximum or minimum of the function, expressed in terms of Δx, the distance from a, is

$$\Delta x = -\frac{f'(a)}{f''(a)} \tag{4.22}$$

It is perhaps easiest to see how this approach works by using a numerical example. Let's start with a function that has a minimum but is clearly not quadratic. A quadratic function would be approximated exactly by a two-term Taylor series, so it wouldn't be very instructive. Consider this non-quadratic function with a minimum at $x^* = 1.37111$.

$$f(x) = e^{-x} + \sqrt{(x-1)^2 + 2} \tag{4.23}$$

The iteration process to locate the minimum of this function is very short. Like most iterative algorithms, it needs a starting point. Let's start at $x = 0$. Table 4.1 shows that the first three iterations converge quickly to the minimum.

Figure 4.33 shows the function with the minimum value along with the initial guess and the first two approximations to the minimum. The minimum is indicated by the plus sign, "+", and x_0, x_1, and x_2 are shown by circles. Note that x_3 is not

Table 4.1 Convergence of second-order approximation

n	x_n	Δx	x_{n+1}
0	0	1.13896	1.13896
1	1.13896	0.21861	1.35758
2	1.35758	0.01348	1.37106

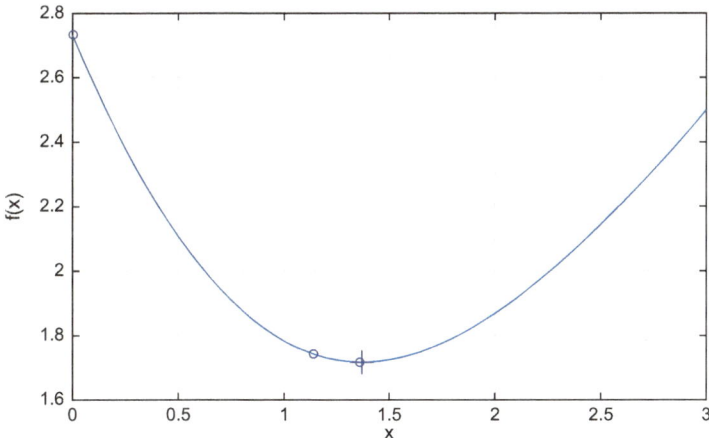

Fig. 4.33 Convergence of the second-order estimation process

shown since it would overlap other symbols. It is clear that the process converges nicely, at least in this case.

If this iterative process works with a single variable, it should work for any number of variables. A basic idea about mathematics is that it doesn't generally care how many dimensions you use; single-variable methods are generally just special cases of more general multivariable methods.

The more general form of the two-term Taylor series is

$$f(x) = f(a + \Delta x) \approx f(a) + \nabla f(a)^T \Delta x + \frac{1}{2} \Delta x^T H(a) \Delta x \qquad (4.24)$$

where a and Δx are both vectors. $H(a)$ is the Hessian matrix evaluated at a. The Hessian is a symmetric matrix that contains all possible second derivatives of the objective function. In the case of two design variables, x and y, the Hessian would be

$$H = \begin{bmatrix} \dfrac{\partial^2 f}{\partial x^2} & \dfrac{\partial^2 f}{\partial x \partial y} \\ \dfrac{\partial^2 f}{\partial x \partial y} & \dfrac{\partial^2 f}{\partial y^2} \end{bmatrix} \qquad (4.25)$$

The general form of the Hessian is

$$H_{r,c} = \frac{\partial^2 f}{\partial x_r \partial x_c} \qquad (4.26)$$

where r is the row number and c is the column number. The Hessian is symmetric since it doesn't matter the order in which the derivatives are calculated.

As before, the goal is to set the derivative of the objective function equal to zero and solve for Δx. The derivative is

$$\nabla f(a) + H(a) \Delta x = 0 \qquad (4.27)$$

And solving for Δx gives

$$\Delta x = -H(a)^{-1} \nabla f(a) \qquad (4.28)$$

It is perhaps not clear why Eq. (4.27) is the derivative of Eq. (4.24). Indeed, this step is sometimes omitted from books and articles on Newton's method.

While not the most compact explanation, working through a two-variable example makes the process clear. Start with the compact form of the two-term Taylor series approximation

$$f(x) \approx f(a) + \nabla f(a)^T \Delta x + \frac{1}{2} \Delta x^T H(a) \Delta x \qquad (4.29)$$

Then expand it, remembering that $H_{12} = H_{21}$

$$f(x) \approx \begin{Bmatrix} f_1 \\ f_2 \end{Bmatrix} + \lfloor \nabla f_1 \ \ \nabla f_2 \rfloor \begin{Bmatrix} \Delta x_1 \\ \Delta x_2 \end{Bmatrix}$$

$$+ \frac{1}{2} \lfloor \Delta x_1 \ \ \Delta x_2 \rfloor \begin{bmatrix} H_{11} & H_{12} \\ H_{12} & H_{22} \end{bmatrix} \begin{Bmatrix} \Delta x_1 \\ \Delta x_2 \end{Bmatrix} \qquad (4.30)$$

Carrying out the multiplications gives

$$f(x) \approx f(a) + \nabla f_1 \Delta x_1 + \nabla f_2 \Delta x_2$$

$$+ \frac{1}{2} H_{11} \Delta x_1^2 + H_{12} \Delta x_1 \Delta x_2 + \frac{1}{2} H_{22} \Delta x_2^2 \qquad (4.31)$$

$$\frac{\partial f}{\partial \Delta_1} = \nabla f_1 + H_{11} \Delta x_1 + H_{22} \Delta x_2$$

$$\frac{\partial f}{\partial \Delta_2} = \nabla f_2 + H_{12} \Delta x_1 + H_{22} \Delta x_2 \qquad (4.32)$$

$$\frac{\partial f}{\partial \Delta x} = \nabla f(a) + H(a) \Delta x \qquad (4.33)$$

The minimum or maximum point occurs when $\partial f / \partial \Delta x_1 = 0$ and $\partial f / \partial \Delta x_2 = 0$.

If the objective function is really quadratic, the Newton's method, which uses a quadratic approximation, will find the minimum in one iteration.

To show how including second derivatives (curvatures) improves convergence, consider the behavior of the steepest descent and Newton's method when finding the minimum of the Rosenbrock banana function. The starting point (0, 0) forces steepest descent to zigzag in order to reach the minimum as shown in Fig. 4.34. While the steepest descent has several more iterations to go before getting near the minimum, Newton's method has essentially found the minimum in three iterations. The minimum point for the banana function is (1, 1), indicated by the "+" sign, and Newton's method has, after three iterations, reached the point (1.002, 1.005). After the fourth iteration, it reaches the point (1, 1) almost exactly. For clarity, the fourth iteration is omitted.

Note that the first iteration for Newton's method in this case is identical to that for the steepest descent. That is because of the choice of starting point. $H(0, 0) = 2[I]$, so the first search direction is determined only by the gradient.

It is tempting to conclude from Fig. 4.34 that Newton's method always follows a path similar to that of the steepest descent. However, the extra information provided by the curvature can greatly affect the path through design space. As an example, consider another two-variable function that is clearly not elliptical

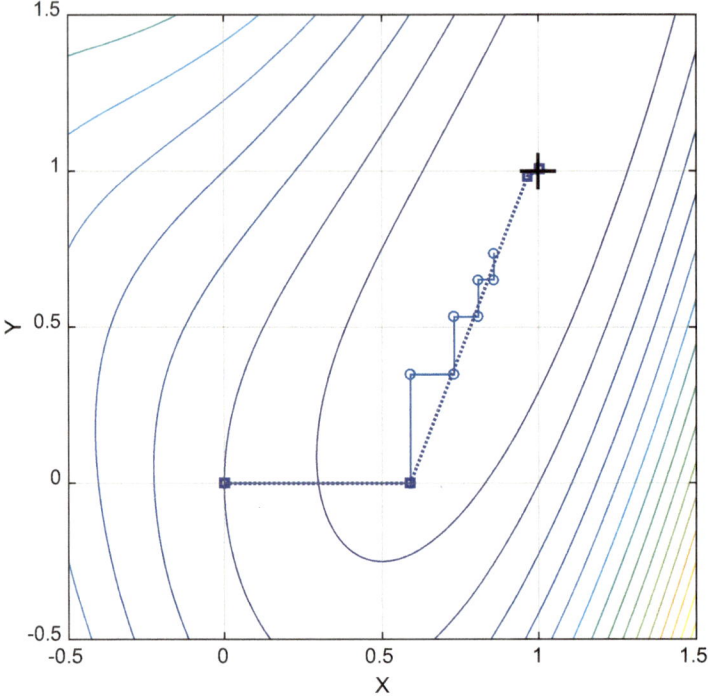

Fig. 4.34 Steepest descent and Newton's method with banana function

$$f(x,y) = x^2 + y^2 + \frac{e^{x+2}}{x^4 + (y-1)^4 + 1} \tag{4.34}$$

The starting point is (1, 1) and we'll look that the paths followed by Newton's method and steepest descent. For the steepest descent method, the first search direction is $S^T = \lfloor 8.043\ -2 \rfloor$ and, as always, all successive search directions are perpendicular to the previous ones. In three iterations, the steepest descent gets close to the minimum point of $(-0.316, -0.662)$. In contrast, Newton's method reaches the minimum in two iterations and does so by taking a very different path (Fig. 4.35).

Remember that Newton's method differs from the steepest descent only in how it calculates the search direction. For the two-variable problem, the updated estimate of the minimum point, (x^*, y^*) is

$$\begin{Bmatrix} x_{k+1} \\ y_{k+1} \end{Bmatrix} = d^* \begin{Bmatrix} S_1^k \\ S_2^k \end{Bmatrix} + \begin{Bmatrix} x_k \\ y_k \end{Bmatrix} \tag{4.35}$$

In the simplest implementation of Newton's method, $d^* = 1$. A modification of this method finds d^* using a 1-D search to find a minimum value of the objective function. In the steepest descent method, d^* is always positive because the search

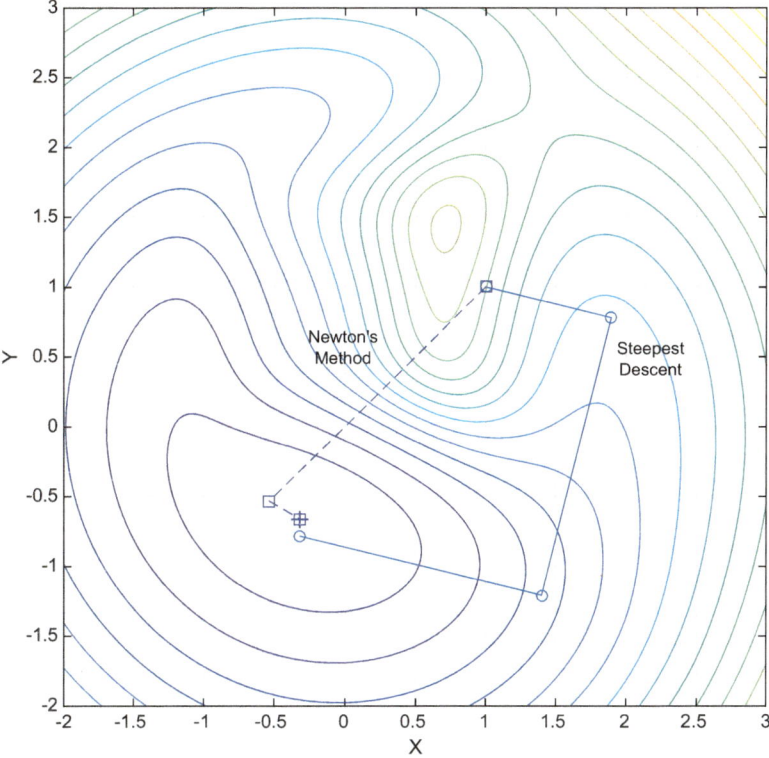

Fig. 4.35 Differing paths taken by the steepest descent and Newton's method

direction is $-\nabla f$. However, S is not necessarily a descent direction in Newton's method. In order to make the plot in Fig. 4.35, d was assumed to be positive. In the first iteration, the closest minimum was actually in the negative search direction and would have put the first estimate of the minimum (x_1, y_1) at (1.566, 1.566). By searching in both positive and negative values of d, the first iteration got close to the minimum value of the objective function. Without searching both directions, the route to the minimum chosen by Newton's method for this example is more circuitous and requires five iterations to find the minimum. It would have then taken one more iteration than the steepest descent.

4.4.6   Quasi-Newton's Methods

The next step in sophistication is quasi-Newton's methods. It is often computationally expensive to calculate the inverse of the Hessian, H^{-1}. As a result, there are a group of methods that replace H^{-1} with an approximation that is easier to compute [7]. The search direction is often described as

$$S = -B^{-1}\nabla f \tag{4.36}$$

where B^{-1} is an approximation to the Hessian that is easier to calculate. Typically, these are identified by the initials of the developers. One of the most popular of the quasi-Newton's methods is designated as BFGS after the developers Broyden, Fletcher, Goldfarb, and Shanno [8]. The approximation to the inverse Hessian used by the BFGS method is

$$B_{k+1}^{-1} = \left(I - \frac{s_k y_k^T}{y_k^T s_k}\right) B_k^{-1} \left(I - \frac{y_k s_k^T}{y_k^T s_k}\right) + \frac{s_k s_k^T}{y_k^T s_k} \tag{4.37}$$

4.5 Approximate Solution to a Differential Equation

Optimization, or more specifically minimization, can be used to find approximate solutions to differential equations. A differential equation is just an equation with a slope as one of its components. The solution to a differential equation is a function that makes that equation true. While not, perhaps, mathematically rigorous, it might be helpful to note that, if the equation contains a slope, the solution is usually a function rather than a number since numbers have slopes of zero. That is, $dC/dx = 0$, where C is a constant (a number).

As a reminder, let's consider a differential equation that is easy to solve and then progress to one that isn't. A familiar example of a differential equation is Newton's law, $F = ma$. Acceleration is the derivative (slope) of velocity, so $a = dv/dt$. A more general form of Newton's law – and something closer to his own version – is $F = m$ (dv/dt).

Imagine a space probe firing its engine to accelerate. The thrust is 30,000 N, and the initial mass of the probe is 1000 kg. However, rocket engines are thirsty, and this one consumes 9.25 kg of fuel per second. If the engine burns for 30 s, find the change in velocity.

The governing equation is $F = m(dv/dt)$ or $dv = (F/m)dt$. F is constant, but m is a function of time, $m(t) = 1000 - 9.25\,t$, so

$$dv = \frac{F}{m(t)} dt = \frac{30000}{1000 - 9.25t} dt \tag{4.38}$$

Integrate both sides of the equation to get

$$\int_{v_0}^{v} dv = \int_{0}^{t} \frac{30000}{1000 - 9.25t} dt \tag{4.39}$$

$$\Delta v = v - v_0 = \int_0^t \frac{30000}{1000 - 9.25t} dt \qquad (4.40)$$

The right-hand side of this expression doesn't have a tidy, closed form solution, but it is easy to use software to carry out the integral numerically, as shown in Fig. 4.36. Note that velocity is almost, but not quite a linear function of time.

Few differential equations are this easy to solve, and there is a whole collection of solution methods for different types. Among the most useful is the one that finds an approximate solution for a type of equation called a boundary value problem. This is an equation where the values of the solution are known at both ends of the solution range. As an example, let's consider an electromagnetism problem [1].

Two concentric metal spheres have radii of 4 cm and 8 cm as shown in Fig. 4.37. The inner sphere has an electric potential of 110 V, and the outer sphere has a potential of 0 V (grounded).

Let's find how the electric potential varies between the spheres if the governing equation is

$$r\frac{d^2v}{dr^2} + 2\frac{dv}{dr} = 0 \qquad (4.41)$$

The exact solution is $v(r) = -110 + 880/r$. The method for finding the exact solution is too involved to be described here. Rather, let's use what we know about

Fig. 4.36 Velocity of space probe

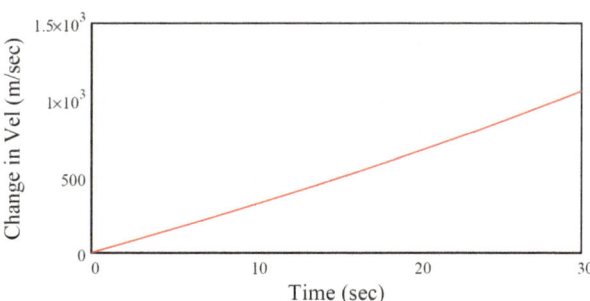

Fig. 4.37 Concentric spheres

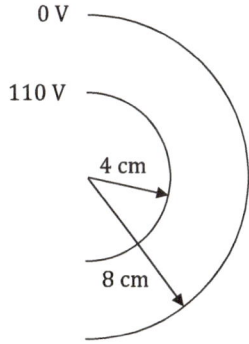

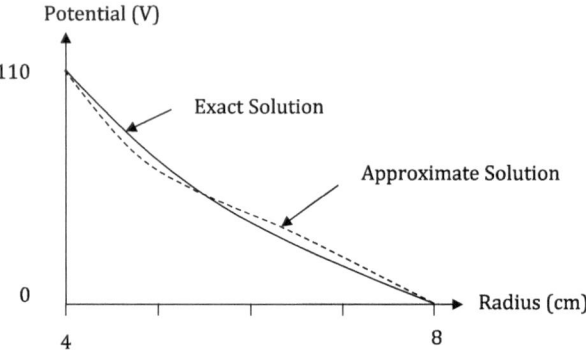

Fig. 4.38 Approximate and exact solutions

optimization to find an approximate solution using an optimization-based approach. For many meaningful problems involving differential equations, it is not possible to find exact solutions, so approximate solutions are the only option. Clearly, the approximate solution should be as close as possible to the exact solution – some measure of the error between the exact and approximate solutions should be minimized.

One way to find an approximate solution is to select a general form of an equation and vary the unknown parameters to minimize the error between the approximate and exact solutions as shown in Fig. 4.38.

The exact solution is the one that goes through both end points and satisfies Eq. (4.41). For an approximate solution, let's guess a third-order polynomial, $v(r) = ar^3 + br^2 + cr + d$. Since this equation doesn't exactly satisfy the governing differential equation, the result will be

$$r\frac{d^2v}{dr^2} + 2\frac{dv}{dr} = \varepsilon(r) \tag{4.42}$$

where $\varepsilon(r)$ is the error due to the approximation. The goal is to find the unknown parameters a, b, c, and d so that the resulting polynomial goes through both end points and minimizes the total error.

The total error is the integral of $\varepsilon(r)$ over the range of interest. We have to ensure that positive errors and negative errors don't cancel one another out. The solution is to square the error so that it is always positive. Then, the only way that the total error can be zero is if $\varepsilon(r) = 0$ everywhere in the range of interest. So, the total error is

$$E = \int_4^8 \varepsilon(r)^2 dr = \int_4^8 \left[r\frac{d^2v}{dr^2} + 2\frac{dv}{dr} \right]^2 dr \tag{4.43}$$

The derivatives of the assumed function are

$$v(r) = ar^3 + br^2 + cr + d$$

$$\frac{dv}{dr} = 3ar^2 + 2br + c \tag{4.44}$$

$$\frac{d^2v}{dr^2} = 6ar + 2b$$

Substituting these expressions into the expression for error gives

$$\varepsilon(r) = r\frac{d^2v}{dr^2} + 2\frac{dv}{dr} = r[6ar + 2b] + 2[3ar^2 + 2br + c] \tag{4.45}$$

Simplifying gives

$$\varepsilon(r) = 12ar^2 + 6br + 2c \tag{4.46}$$

Before solving for the unknown parameters, we need to ensure that the resulting curve goes through both end points, $v(4) = 110$ and $v(8) = 0$.

$$v(4) = a(4^3) + b(4^2) + c(4) + d = 110$$
$$v(8) = a(8^3) + b(8^2) + c(8) + d = 0 \tag{4.47}$$

Solving for the two end conditions will eliminate two of the unknowns, and the remaining two will be design variables for the error minimization problem. There are different ways to substitute the end conditions into the expression for error and all are mathematically equivalent. If you choose a different, but valid, solution, that's fine.

The first of the two end point conditions can be used to solve for c so that

$$c = 27.5 - 16a - 4b - 0.25d \tag{4.48}$$

And the second can be solved for d so that

$$d = -512a - 64b - 8c \tag{4.49}$$

Substituting the second of these expressions into the first gives

$$c = -27.5 - 112a - 12b \tag{4.50}$$

Finally, substituting back into Eq. (4.46) gives

$$\varepsilon(r) = (12r^2 - 224)a + (6r - 24)b - 55 \tag{4.51}$$

So the total error is

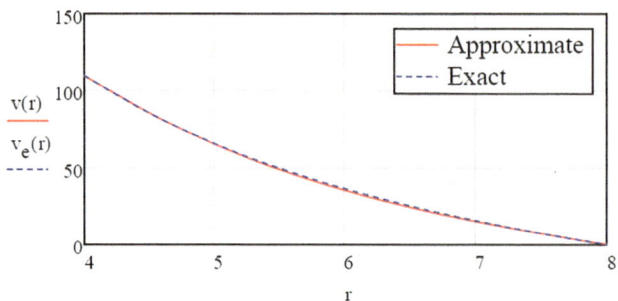

$$E(a,b) := \int_4^8 \left[\left(12 \cdot r^2 - 224\right) \cdot a + (6r - 24)b - 55\right]^2 dr \qquad \text{<= Define total error expression}$$

$$a := 0 \qquad b := 0$$

$$\binom{a}{b} := \text{minimize}(E, a, b) = \binom{-0.716}{17.76} \qquad \text{<= Find a and b to minimize error}$$

$$c := -27.5 - 112a - 12 \cdot b = -160.417 \qquad \text{<= Back solve for c and d}$$

$$d := -512 \cdot a - 64 \cdot b - 8 \cdot c = 513.333$$

$$v(r) := a \cdot r^3 + b \cdot r^2 + c \cdot r + d \qquad v_e(r) := -110 + \frac{880}{r} \qquad \begin{array}{l}\text{<= Compare approximate and}\\ \text{exact solutions}\end{array}$$

Fig. 4.39 Approximate and exact solutions

$$E = \int_4^8 \varepsilon(r)^2 dr = \int_4^8 \left[\left(12r^2 - 224\right)a + (6r - 24)b - 55\right]^2 dr \qquad (4.52)$$

Figure 4.39 shows the Mathcad computation and a comparison with the exact solution.

4.6   Evolution-Inspired Semi-Random Search: Following Nature's Lead

As mentioned in the first part of this chapter, evolution is one of nature's forms of optimization. This section shows an evolution-inspired algorithm similar to the one described by Stephen Jones.

It's easy to imagine that there are numerous algorithms that mimic the behavior of evolution, and, indeed, the field of genetic algorithms is an active area of research [9]. A representative genetic algorithm encodes each individual as a finite length vector of components in some alphabet, often [0, 1]. These vectors are analogous to chromosomes, and the individual components are analogous to genes. Genetic algorithms select individuals who maximize their fitness – the extent to which they maximize or minimize an objective function – through processes that modify, pass on, and combine these chromosomes. The method described here is a simple one inspired by evolution, but doesn't rise to the level of a genetic algorithm. This approach might be called an evolutionary strategy.

To make the operation clear, let's start with a reasonably simple objective function

$$z = \frac{x+y+1}{x^2+y^2+1} + \frac{x^2+y^2}{100} \tag{4.53}$$

Figure 4.40 shows this function in 3D. The minimum occurs at $x = -1.238$, $y = -1.238$.

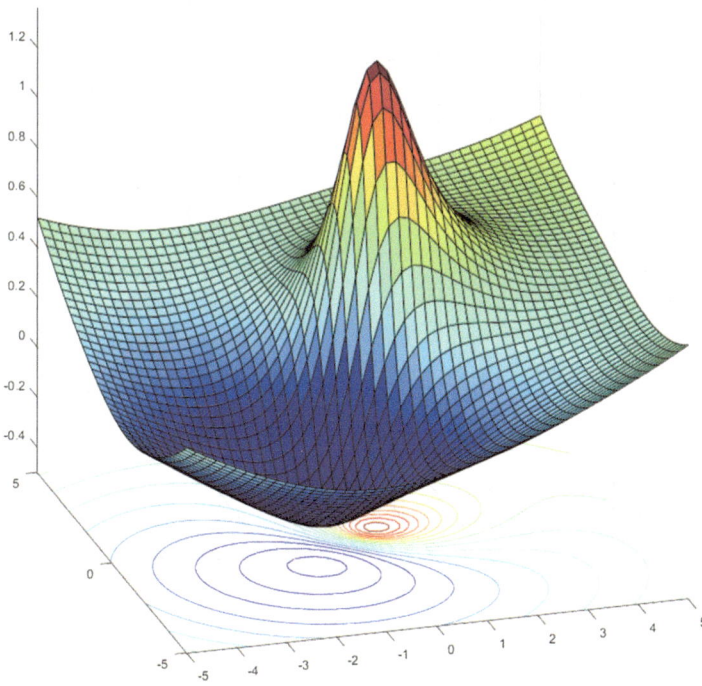

Fig. 4.40 Test function for evolutionary algorithm example

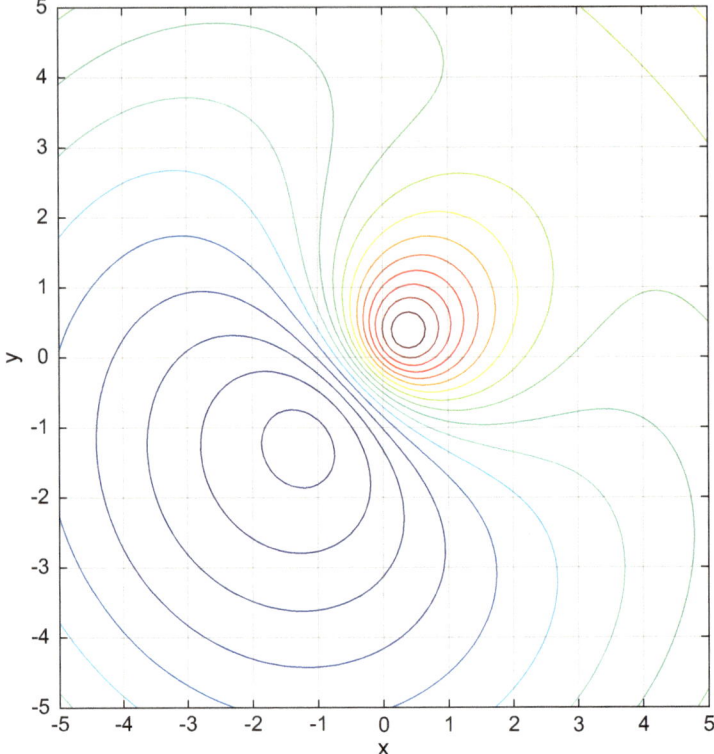

Fig. 4.41 Contour plot of evolution algorithm test function

Figure 4.41 shows a contour plot of the test function. It's clear that there are starting points that would require a circuitous route to the minimum.

If, for example, the starting point were $x = 0$, $y = -4$, the search direction for the steepest descent method would point very nearly at the minimum. If, in contrast, the starting point was $x = 1$, $y = 2$, the search direction would point well away from the minimum. Indeed, the search path would need to arc around the peak of the objective function at $x = 0.369$, $y = 0.369$.

Figure 4.42 shows the first iteration or generation perhaps. The box shows the starting point, and the circles show the randomly chosen points where objective function values are calculated. Finally, the diamond shows the minimum – the point the algorithm is trying to find. There are ten random points calculated for each iteration. There needs to be some range defined for x and y. For this example, both were allowed to vary by ± 2 from the starting point. The minimum point is the one at the upper left.

Figure 4.43 shows the second iteration. The starting point is the minimum of the randomly chosen points from the previous iteration. Again a cloud of ten randomly chosen points surrounds the starting point. The point corresponding to the lowest objective function value is the one farthest to the left, $x = -1.93$, $y = 3.786$.

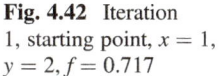

Fig. 4.42 Iteration
1, starting point, $x = 1$,
$y = 2, f = 0.717$

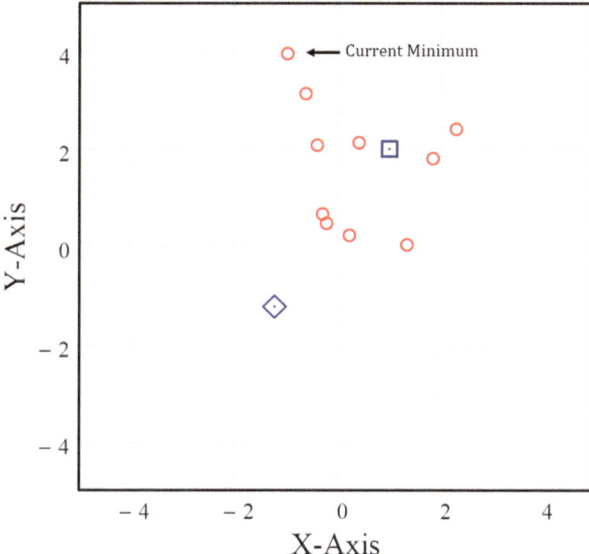

Fig. 4.43 Iteration
2, starting point,
$x = -0.995, y = 3.954$,
$f = 0.391$

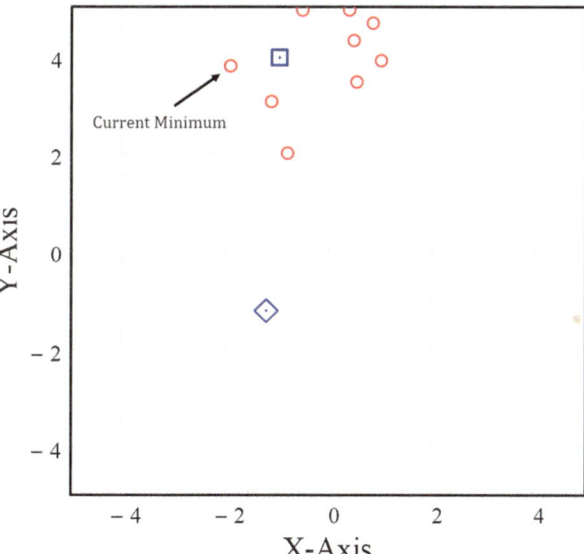

It is worth tracking one more iteration before looking at the overall performance
of the algorithm. Figure 4.44 shows the third iteration, with behavior typical of that
shown in the previous iterations. It appears now that the solution path has gotten
around the peak in the objective function and will likely progress smoothly to the
minimum.

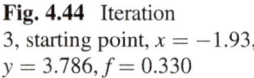

Fig. 4.44 Iteration
3, starting point, $x = -1.93$,
$y = 3.786, f = 0.330$

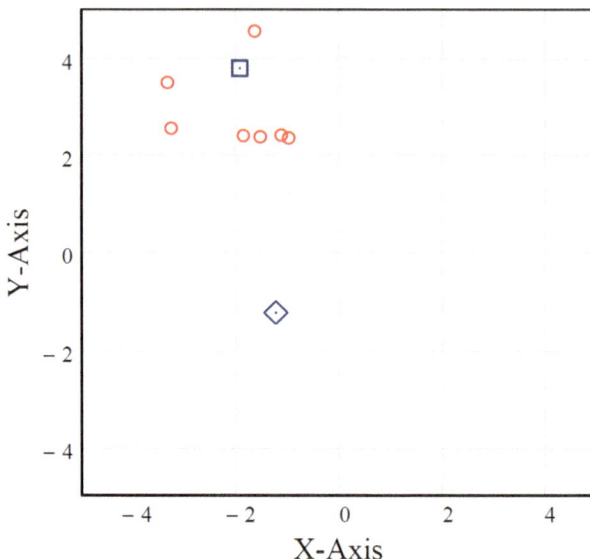

Fig. 4.45 Iteration
7, starting point,
$x = -1.358, y = -1.08$,
$f = -0.328$

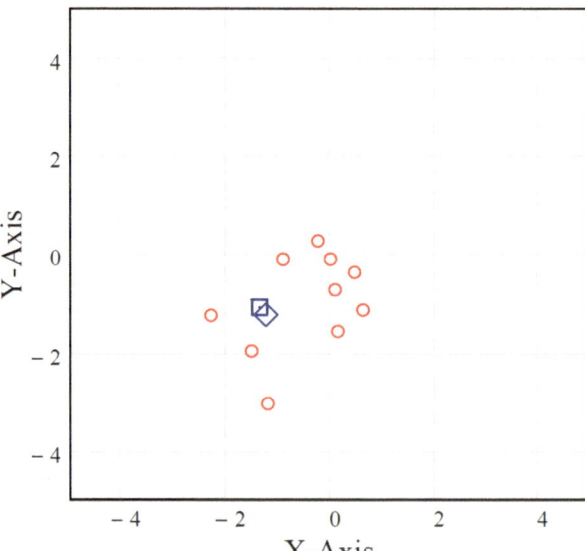

Figure 4.45 shows the result of the seventh iteration. The algorithm has gotten very near the minimum, and it would now require that one of the random points be even closer to the minimum for there to be any improvement to the estimated minimum. In fact, the eighth iteration gave a worse estimate for the minimum than does the seventh iteration.

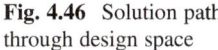

Fig. 4.46 Solution path through design space

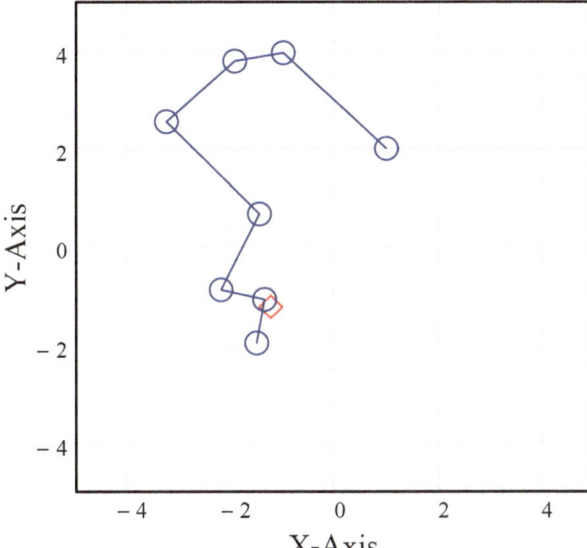

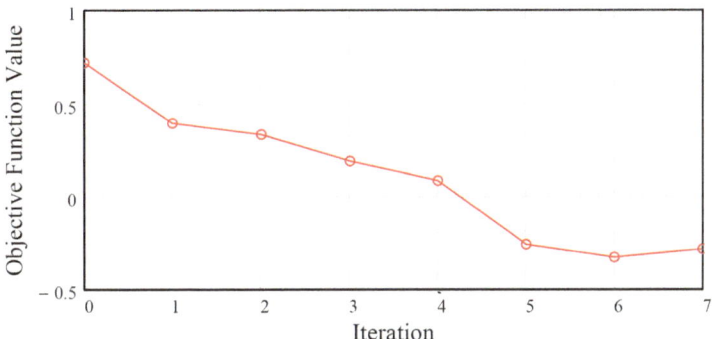

Fig. 4.47 Convergence plot

Figure 4.46 shows the path taken by the algorithm through design space. Again, the diamond shows the point corresponding to the minimum value of the objective function. It's clear that the algorithm got very near the minimum, but continued for one more iteration. In practice, this might satisfy one of the exit criteria.

Finally, Fig. 4.47 shows a convergence plot of the objective function vs. iteration number.

There are several possible exit criteria for this approach:

- Stop when the next iteration gives an increase in the objective function.
- Use more random points for the last iteration in the hope that one of them will be closer to the minimum than the starting point.
- When the objective function increases, reduce the allowable variation in the design variables, Δx and Δy in this case, and repeat the last iteration.

Select $x_1^0, x_2^0, \cdots, x_n^0, \Delta x_{min}, \Delta F_{min}, M_{max}$

m=0

Select range for random variables, d

Select number of random points, N

Generate random points: $rx_1^1, rx_2^1, \ldots, rx_n^1, \ldots, rx_1^N, rx_2^N, \ldots, rx_n^N$ such that

$$x_1^0 - d < rx_1^0 < x_1^0 + d, \cdots$$

Calculate F at each point: $F(rx_1^1, rx_2^1, \ldots), F(rx_1^2, rx_2^2, \ldots), \ldots$

Find F* and {rx*}

$$x_1^0 = rx_1^{*1}, \ldots$$

Exit Criteria satisfied?

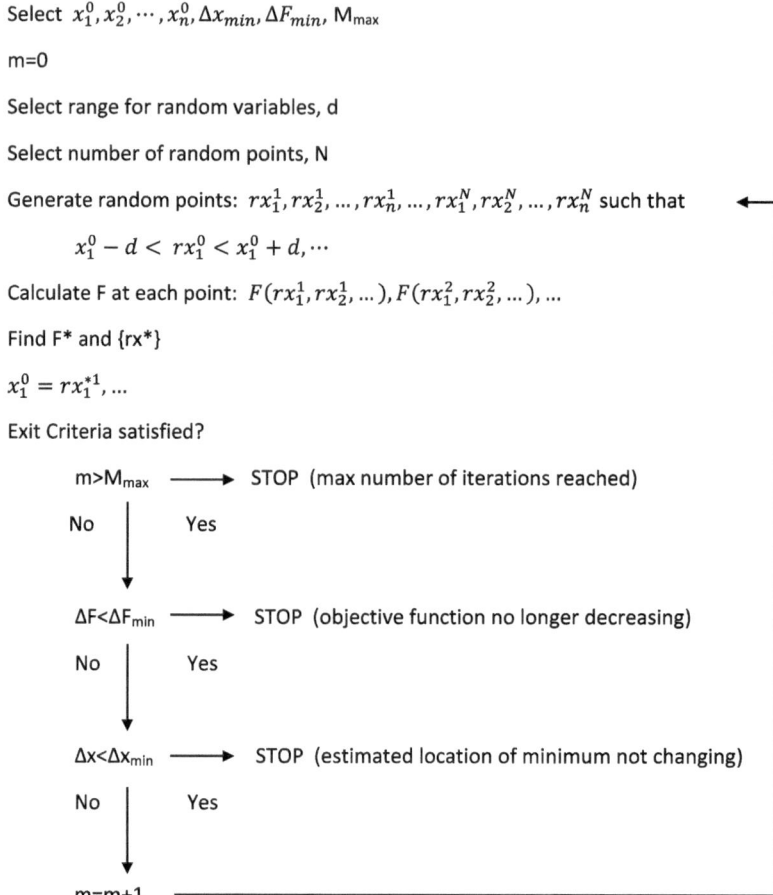

$m > M_{max}$ $\longrightarrow$	STOP (max number of iterations reached)
No Yes	
$\Delta F < \Delta F_{min}$ $\longrightarrow$	STOP (objective function no longer decreasing)
No Yes	
$\Delta x < \Delta x_{min}$ $\longrightarrow$	STOP (estimated location of minimum not changing)
No Yes	
m=m+1	

Fig. 4.48 Pseudo flow chart for evolution-inspired algorithm

- Adopt a hybrid approach and use some other method, such as a surface fit, to further refine the estimate of the minimum based on the last set of randomly chosen points.

Figure 4.48 shows a pseudo flow chart for a simple implementation of this method. Note that this is the simplest implementation of the method. A more refined version might reduce the range, d, when Δx became small.

4.7 Convex Problems

Many of the practical problems in optimization fade if the problem is convex. For an unconstrained problem, this just means that the feasible region of the objective function is convex. Graphically speaking, a convex function is one in which line

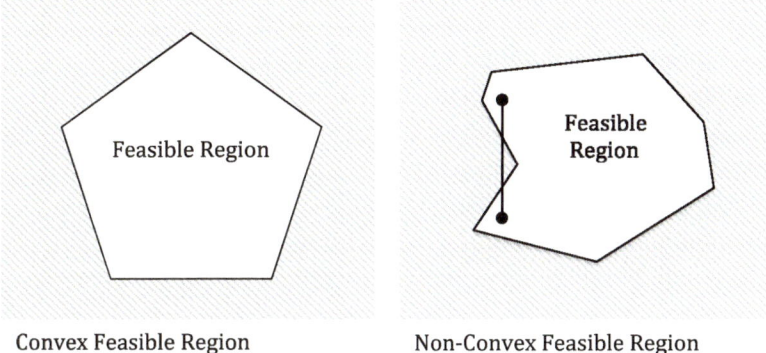

Convex Feasible Region   Non-Convex Feasible Region

Fig. 4.49 Examples of convex and non-convex shapes

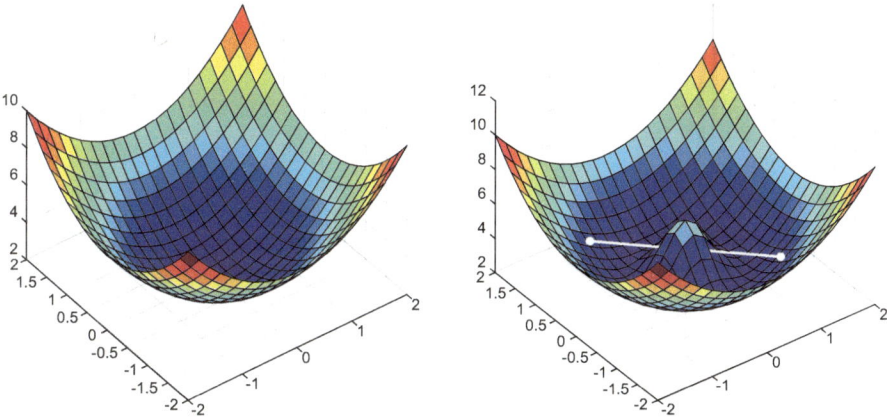

Fig. 4.50 Convex and a non-convex objective functions

segments connecting any two points in the feasible region are, themselves, inside the feasible region. Figure 4.49 shows both convex and non-convex shapes.

In three dimensions, a simple elliptical function is convex, as shown in Fig. 4.50. The surface on the left side is a simple elliptical function, $z = x^2 + y^2$, and is convex. Note that a convex function is one for which the Hessian is positive semidefinite. The objective function on the right is clearly non-convex since a line segment joining two points in the feasible region leaves the feasible region.

There is one obvious thing to note about these pictures of convex functions: any engineer would naturally call them concave. A bowl, for example, is concave, not convex, and the shapes of convex and concave lenses are well-known to anyone who wears glasses. Whether they are intuitively convex or concave depends completely on one's point of view - for example, whether you are looking at the outside or inside of the bowl.

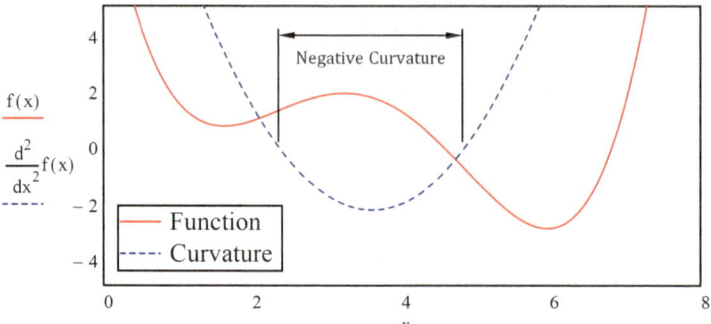

Fig. 4.51 Negative curvature between minima

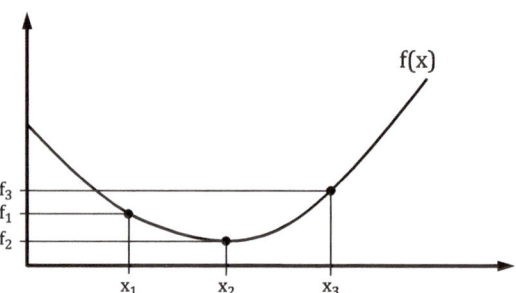

Fig. 4.52 Selected points
on a convex function

The surfaces shown in Fig. 4.50 above are being viewed from the top and appear to be concave. However, it's just as valid to view them from the bottom, where they would appear to be convex. It is for reasons like this one that mathematicians are careful to define concepts in clear, compact, and unambiguous terms. The physical understanding of problems that often mark the intellect of a good engineer may be much less useful to a mathematician working with more abstract concepts.

The point of worrying about whether a function is convex or not is what it implies for optimization problems. A convex function has a single global minimum. Put another way, two local minima are necessarily separated by a non-convex region. Figure 4.51 shows a simple example of a function with one local and one global minimum.

The descriptions of convexity so far have been graphical, but it's important to have a more precise definition [10]. A function of a single variable, $f(x)$, is convex over an interval $[a, b]$ if its slope increases as x increases. Given any three points on the interval, such that $x_1 < x_2 < x_3$, the function is convex if

$$\frac{f_2 - f_1}{x_2 - x_1} \leq \frac{f_3 - f_2}{x_3 - x_2} \tag{4.54}$$

As shown in Fig. 4.52, this idea is easy to extend to more dimensions.

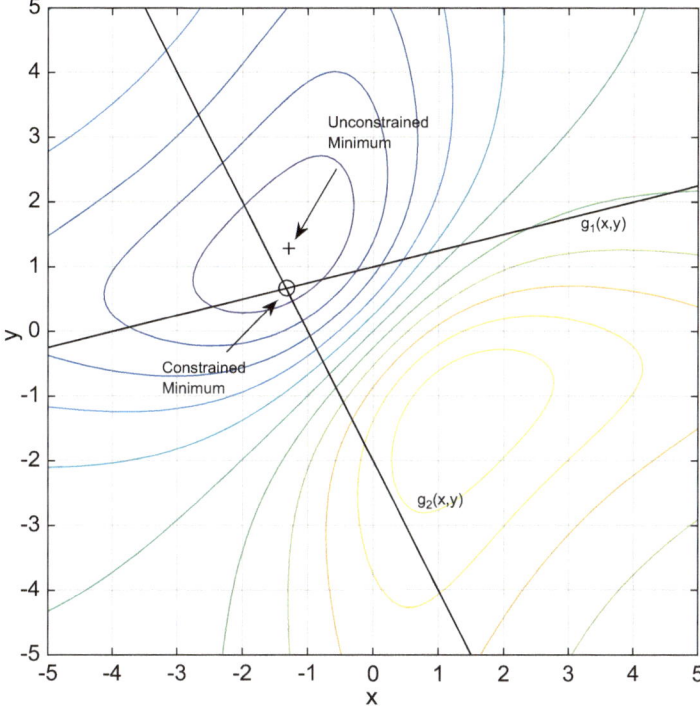

Fig. 4.53 Constrained butterfly function

An equivalent definition is that a function is convex if its Hessian is positive semidefinite.

This is probably the time for a brief digression on some classes of optimization methods. References on convex optimization [11] usually talk about a class of algorithms called interior point methods. This distinction is useful because some optimization methods move through the interior volume of design space, while others move along the surface that is the boundary between the feasible and infeasible regions. For example, the steepest descent and Newton's method are interior point methods. Interior point methods are often used for convex problems.

We know that convex functions have a single global minimum, and we have some workable definitions of convex functions. The remaining hurdle is that few engineering problems are convex. Thus, there needs to be a way to approximate non-convex problems as convex ones, perhaps successively as part of an algorithm. Fortunately, such methods exist and are part of some commercial software packages. One of these methods is sequential linear programming (SLP).

4.7.1   Sequential Linear Programming

Linear problems, whose objective functions and constraints are all linear functions of the design variables, are necessarily linear. So, if a nonlinear problem can be expressed approximately as a series of linear approximations, those approximate problems can be solved relatively easily. SLP iteratively finds the minimum of a constrained nonlinear problem by linearizing the problem, solving the now convex linear problem to estimate the minimum and then re-linearizing about the new point. This process is repeated until the solution converges.

As an example, consider the butterfly test function from Appendix B, with two linear constraints added:

$$\text{Minimize } f(x, y) = \frac{x - y}{(x^2 + 5)(y^2 + 5)} + \frac{y^2}{20000}$$

$$\text{Subject to } g(x, y) = y - \frac{1}{4} - 1 \leq 0$$

$$g_2(x, y) = -y - 2x - 2 \leq 0$$

The design space for this problem is shown in Fig. 4.53, including both the constrained and unconstrained minima.

Linearizing an equation means replacing the nonlinear expression with its one term Taylor series approximation. The general form is

$$f(x) \approx f(a) + (x - a)^T \nabla f(a) \tag{4.55}$$

Where both x and a are assumed to be vectors

For the two-variable objective function, the general expression, expanded at the point $[a, b]$, becomes

$$f(x, y) \approx f(a, b) + \lfloor (x - a) \quad (y - b) \rfloor \left\{ \begin{array}{c} \frac{\partial f}{\partial x}\big|_{a,b} \\ \frac{\partial f}{\partial a}\big|_{a,b} \end{array} \right\}$$

$$\approx f(a, b) + \frac{\partial f}{\partial x}\bigg|_{a,b}(x - a) + \frac{\partial f}{\partial a}\bigg|_{a,b}(y - b) \tag{4.56}$$

The constraints are already linear, so they don't need to be modified. In the more general case, linearized expressions for the constraints follow the same form as shown in Eq. (4.56). Figure 4.54 shows the details of the calculation. Note that, for this example, the linear approximation finds the constrained minimum in a single iteration, though this isn't generally the case. It is also interesting to note that the starting point was outside the feasible region, showing that this method can have an infeasible starting point.

The point of SLP is that linear problems are very easy to solve, in part because they are convex. Thus, solving a succession of linear approximate problems might be more efficient than a method that works directly on the nonlinear problem. One possible annoyance is boundedness. Not all linear problems have a solution, and those that don't are called unbounded. The minimum of a linear constrained problem

Define objective function and its derivatives

$$f(x,y) := \frac{x - y}{\left(x^2 + 5\right)\cdot\left(y^2 + 5\right)} + \frac{y^2}{20000}$$

$$dfdx(x,y) := \frac{d}{dx}f(x,y) \rightarrow \frac{1}{\left(x^2 + 5\right)\cdot\left(y^2 + 5\right)} - \frac{2\cdot x\cdot(x - y)}{\left(x^2 + 5\right)^2\cdot\left(y^2 + 5\right)}$$

$$dfdy(x,y) := \frac{d}{dy}f(x,y) \rightarrow \frac{y}{10000} - \frac{1}{\left(x^2 + 5\right)\cdot\left(y^2 + 5\right)} - \frac{2\cdot y\cdot(x - y)}{\left(x^2 + 5\right)\cdot\left(y^2 + 5\right)^2}$$

Define constraint functions and derivatives $g_1(x,y) := y - \dfrac{x}{4} - 1$ $g_2(x,y) := -y - 2\cdot x - 2$

Find the constrained minimum of nonlinear problem $x := 0$ $y := 0$ <= Initial guess

Given $g_1(x,y) \le 0$ $g_2(x,y) \le 0$ $\text{Minimize}(f,x,y) = \begin{pmatrix} -1.333 \\ 0.667 \end{pmatrix}$

Define the linearized functions at the starting point, x=a, y= $a := 1$ $b := 2$

$g_1(a,b) = 0.75$ <= Infeasible starting point $dfdx(a,b) = 0.025$

$g_2(a,b) = -6$ $dfdy(a,b) = -0.01$

$$f_{lin}(x,y) := f(a,b) + dfdx(a,b)\cdot(x - a) + dfdy(a,b)\cdot(y - b) \rightarrow \frac{2\cdot x}{81} - \frac{12257\cdot y}{1215000} - \frac{27743}{1215000}$$

Find the minimum of the linearized constrained problem

Given

$g_1(x,y) \le 0$ $g_2(x,y) \le 0$ $\text{Minimize}(f_{lin},x,y) = \begin{pmatrix} -1.333 \\ 0.667 \end{pmatrix}$

Fig. 4.54 Linearized solution for the constrained butterfly problem

lies either along one constraint boundary or at the intersection of two or more constraint boundaries as shown in Fig. 4.55. If the gradient pointed the other way, both of these examples would be inbounded – there would be no limit to the direction one could move in the $-\nabla F$ direction.

4.8 Optimization at the Limits: Unlimited Class Air Racers

Having just gone through numerical problems with clear definitions, it might be helpful to consider an activity in which optimization is at the very core, but one in which the problem is not always well defined – a problem in which millions of dollars of equipment are on the line and the safety of human beings is at risk.

One of the more captivating motor sports in the USA is air racing [12]. Every year, pilots, ground crews, and spectators gather in Reno NV to race airplanes at very low altitudes around a closed course. Before the recent addition of jets, the fastest

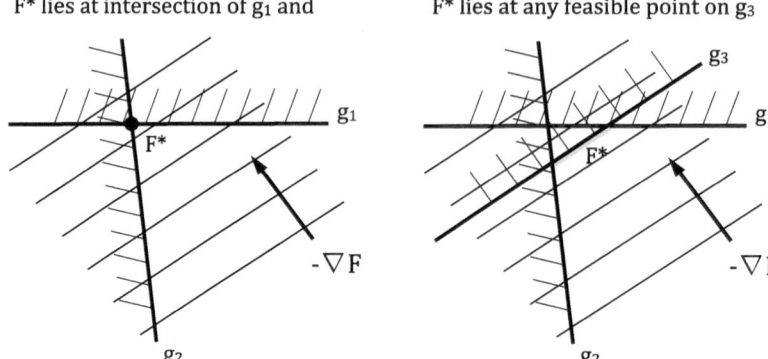

Fig. 4.55 Two possible solutions for a linear problem

and most glamorous planes flew in the unlimited category. The development of these planes is a great example of optimization applied to existing products. First, some background.

The unlimited course at Reno is roughly an oval 13.01 km (8.085 mi) long. The distance is only approximate since it is calculated assuming a speed of 807 km/h (500 mi/h) and turns at 4 g. Planes must complete six laps, and the first one across the finish line wins. Minimum race altitude is 15.2 m (50 ft) above ground level, and the maximum race altitude is generally 76.2 m (250 ft). This is pretty crazy stuff.

The name unlimited suggests that there are no rules governing the planes themselves, and this is very nearly true. They must have piston engines driving propellers, but the planes are otherwise unrestricted [13]. The most pressing limitation is posed by the aerodynamics involved in trying to fly propeller driven planes close to the speed of sound.

The task is then to build a plane that flies as fast as possible at low altitude around a short closed course. The objective function is the time required to complete six laps. There are many constraints, but builders have freedom not usually afforded to people designing airplanes.

A racing plane doesn't need to fly very far or to carry any payload apart from the fuel and pilot. It doesn't have to be easy to handle since only highly skilled race pilots will ever fly it. It doesn't even have to be very good at turning right. It just needs to go fast and turn left for six laps (imagine NASCAR in the air). It can, thus, be highly optimized. So what solutions to this optimization problem have people arrived at?

Almost exclusively, unlimited racers are highly modified WWII fighters. They often have clipped wings and small, streamlined canopies to reduce drag. They also have dramatically more powerful engines than were originally installed. Over the last decade or so, there have been basically two groups of unlimited racers, Mustangs and radials.

The Mustangs started out as North American P-51 Mustang fighters and have been subject to heavy modification. They use liquid-cooled V-12 inline engines –

Fig. 4.56   An unlimited air racer, a modified P-51 named "Voodoo" (Wikimedia Commons, image is in the public domain)

usually Rolls-Royce Merlins – though modified to produce more than the original power. These inline engines have a relatively small cross-sectional area that allows the plane to be narrow and streamlined. However, in order to make the required power, they are run very close to their structural limits. Indeed, it is not unheard of for an engine to fail during a race because of overstress. In this case, the pilot must declare an emergency and either land the plane without power or bail out. Figure 4.56 shows a modified P-51 called Voodoo, a very successful unlimited class racer.

The other group of racers consists of planes powered by large, air-cooled radial engines. This group is populated largely by Grumman F8F Bearcats and Hawker Sea Furies. The typical engine is very large, with as many as 28 cylinders. For example, the Wright R-3350 has a displacement of 54.86 L (3347 cu. In.) and the Pratt and Whitney R-4360 Wasp Major has a displacement of 71.5 L (4362 cu. In.). They are also very powerful, with the Wasp Major probably making more 2980 kW (4000 hp). For obvious reasons, it's difficult to test one of these engines on a dynamometer to do a direct measurement, so power is necessarily an estimate.

One of the most well-known of the radial engine racers is a highly modified Grumman Bearcat called Rare Bear (Fig. 4.57). It has been very competitive over the years and holds the world speed record for piston powered planes at 850.25 km/h (528.33 mi/h). It uses a very large, 18 cylinder, air-cooled radial engine (a Wright R-3350) with much more frontal area than an inline, liquid-cooled engine. The assumption is that the huge power is worth the additional aerodynamic drag. Carefully-designed cowling around the engine also minimizes drag; adding engine

Fig. 4.57 The highly modified Grumman F8F Bearcat, Rare Bear. (Wikimedia Commons, image is in the public domain)

heat to the air flow around the engine adds energy and can significantly reduce drag. Because of the heat added to the flow, it is possible for cooling systems to generate thrust.

These planes have been stripped of all unnecessary equipment, and what systems remain are often highly optimized. In at least one case, the hydraulic mechanism that raises the landing gear was replaced by a lighter system built around a small cylinder of compressed gas. It could raise the landing gear only once before needing to be recharged, but that was enough; there is no reason to raise the landing gear twice in one race.

An obvious question is why unlimited racers are essentially all made from piston engine fighters, the newest of which are now about 70 years old. There are basically two approaches to making an unlimited racer. The first is to start with an existing, proven airplane and modify it. The second is to start from scratch and design a plane for the sole purpose of air racing. The second approach would seem to be the one that would result in the most optimized design, but such planes are extremely rare. There must be a reason that the most successful racers are all converted fighters.

Designing and building from scratch a plane intended to fly at the absolute limits of the aerodynamically possible is not at all easy. The conceptual design might not be prohibitively difficult, but the details of structural and aerodynamic analysis are time-consuming and expensive. Then, there would need to be static structural tests and flight tests of prototypes, probably resulting in design changes, before a race-ready plane could be built. The result would be a highly optimized airplane, but the resources needed are likely to be prohibitive. One constraint that cannot be ignored is budget. There have been only a few attempts at purpose-built unlimited racers, and none have been very successful. No custom-built plane has ever won in the gold category in the unlimited class.

It is worth noting that, in slower classes where the demands are more modest, custom-built designs are dominant, if not universal. However, because of cost limitations, the unlimited class consists exclusively of production aircraft whose development costs were amortized long ago.

The fastest production piston-driven planes were single-seat fighters dating from the end of the WWII, so these would seem to be a good starting point. They were designed to have strong structures, to be fast, to be reliable, and to have handling characteristics within the capabilities of an average, trained pilot. On the face of it, an unmodified piston engine fighter would seem to be an easy choice for an unlimited class racer; however, it's not quite that simple.

A piston engine fighter typically had many different possible uses, so raw speed was not the only priority. The result was invariably a plane that was good at a number of things, but not optimized for any one mission. Thus, a race plane builder needs to develop a plane as optimized for racing as it can be, but starting with a plane that had to satisfy many requirements and was the result of many compromises.

For example, the original planes needed to be survivable, so they were fitted with armor plating and heavy, self-sealing fuel tanks. They had to be well armed, so they were fitted with heavy machine guns in each wing along with storage for all the ammunition they consumed. Pilots needed to see what was around them, so most designs were fitted with large, bubble canopies. The engines were large and powerful, but not so much that range was compromised. The list is a long one that speaks of many design choices between conflicting requirements.

Think now of turning a fighter into a racer. In optimization terms, this means starting with a plane that had no single objective function and turning it into a plane with a single purpose – one clearly defined objective function. Because there are limits to what can be changed about the aircraft, the result is necessarily as optimized as it might be.

Things that can be changed include limited weight reduction by removing components that are no longer needed. Also, the engine no longer needs to be reliable over long distances or to have a long service life, so it can be modified to increase power for a short time. Alternately, a larger, heavier, and more powerful engine can be substituted.

However, some things just can't realistically be changed. The basic structure would be extremely difficult to modify beyond simple reinforcements. Also, the overall aerodynamic envelope would be difficult to modify without affecting, perhaps badly, the behavior of the plane in the air. Thus, the number of modifications – the extent to which a plane can be optimized for its new role as a racer – is limited.

There has been at least one attempt to drastically change the aerodynamics of an old fighter plane. In the 1990s, a P-51D was fitted with the wings and horizontal tail from a Lear Jet (a small business jet). The resulting plane was dubbed Miss Ashley II. It showed promise in its early races, but in 1999, the plane crashed, killing the pilot. Witnesses saw the right wing fold back in the first turn, though it is possible that wasn't the initial cause of the accident. One failed attempt is not enough evidence to conclude that drastic aerodynamic modifications are a bad idea, but it is enough to serve as a warning that these kinds of changes are not to be made casually.

Here's an incomplete list of common modifications done to fighters to make them race planes – to optimize them for a single task:

- Modify engine or replace with larger, more powerful engine.
- Clip the wings for decreased drag at high speed.
- Remove all armament and armor.
- Replace bubble canopy with smaller, streamlined canopy.
- Remove original equipment like heavy tube radios.
- Modify or replace secondary internal systems to save weight.
- Increase size of the vertical tail to compensate for increased torque.

The result can be a plane capable of 805 km/h (500 mi/h) laps at Reno. But it will definitely be something of a force fit. The plane, while more optimized for its role that it originally was, is still basically a 70 year old fighter. It was designed for a different mission, using slide rules and drafting boards, by people who were very smart, but knew nothing of modern aerodynamic and structural analysis.

That plane will also be eye-wateringly expensive. Just the plane with which you start may now cost millions. The engine will require high octane aviation fuel and will gulp the stuff in vast quantities. The parts you will need mostly haven't been made for at least 60 years, so you may find yourself paying high prices for needed components or improvising by using what parts are available. In short, your plane might be the stuff of which bankruptcies are made.

It is interesting to note that the fastest unlimited races are not the most recent ones. The rules remain unchanged, so one might guess that winning speeds over the years would have steadily increased. However, it seems that teams' abilities to field increasingly rare airplanes maintained by increasingly rare parts is a variable thing.

As this was written, the unlimited class race record at Reno was set in 2003 by Skip Holm, flying a heavily modified P-51D named (unfortunately perhaps) Dago Red. The average speed during the race was 816.106 km/h (507.105 mi/h). However, in 2014, the most recent results as this was written, the winner was Steve Hinton, flying a modified P-51D named Voodoo. The winning speed was "only" 792.642 km/h (492.525 mi/h) – fast, but noticeably slower than the 11-year-old record. In 2014, Rare Bear, then holder of the world piston speed record, came in second with a race speed of 767.462 km/h (476.879 mi/h). There are many possible reasons for this variability, but money is a big part. A famous observation from car racing comes to mind: Speed costs money. How fast do you want to go?

In the interest of completeness, as this is being written, the fastest custom-built (also called homebuilt) racer is the Nemesis NXT, designed by Jon Sharp. Several examples are flying, and they are very competitive in a class called super sport – a step down from unlimited. The Nemesis NXT uses modern materials and aerodynamics, and Sharp set a super sport race record of 655.1 km/h (407.1 mph).

Fig. 4.58 Jon Sharp's Nemesis NXT

Perhaps more impressively, Sharp set that record with a production light aircraft engine that is nominally rated at 261 kW (350 hp). Thus, he raced at almost unlimited class speeds (over 400 mph has often been considered the dividing line) with much, much less power. Clearly his design is much more optimized for its task and more efficient. Figure 4.58 shows Nemesis NXT in flight.

References

1. Kreszig E (2011) Advanced engineering mathematics, 10th edn. Wiley, Hoboken
2. Hamming R (1987) Numerical methods for scientists and engineers. Dover, Mineola
3. Sendeckyj G (1990) Interviewee, marching grid algorithm. [interview]. Dayton
4. Curry HB (1944) The method of steepest descent for non-linear minimizaton problems. Q Appl Math 2(3):258–261
5. Fletcher R, Reeves C (1964) Function minimization by conjugate gradients. Comput J 7 (2):149–154
6. Kelley C (1999) Iterative methods for optimization, Society for Industrial and. Appl Math, Philadelphia
7. Fletcher R, Powell M (1963) A rapidly convergent descent method for optimization. Comput J 6 (2):163–168
8. Nazareth L (1979) A relationship between the BFGS and conjugate gradient algorithms and its implications for new algorithms. SIAM J Numer Anal 16(5):794–800
9. Gen M, Cheng R (1997) Genetic algorithms and engineering design. Wiley, Hoboken
10. Eggleton RB, Guy RK (1988) Catalan strikes again! How likely is a function to be convex? Math Mag 61:211–219

11. Boyd S, Vandenberghe L (2004) Convex optimization. Cambridge University Press, Cambridge
12. Handleman P (2007) Air racing over Reno: the World's fastest motor sport. Specialty Press, Forest Lake
13. Reno Air Racing Association, "Reno Championship Air Races," [Online]. Available: http://airrace.org/. Accessed 16 June 2016

Chapter 5
Constraints: Placing Limits on the Solution

A problem well stated is a problem half solved

-Charles Kettering

So far, we have talked about problems in which the design variables can take any values they like. In the two-variable lifeguard problems, it wouldn't make sense for the two design variables to assume values far away from about 85 m, but there is no mathematical reason they could not. In practical problems, there are usually limitations on the values that can be assumed by the design variables. These limitations are called constraints.

5.1 Maximum Volume of a Box

A very simple constrained optimization problem is to find the maximum volume box that can be made from a piece of cardboard with a given size. Let's say that the sheet of cardboard is 500 mm wide and 1000 mm tall and that the faces of the box are laid out as shown in Fig. 5.1.

The volume of the box is $V = A \times B \times C$. In addition, there are two constraints. The first is that $2C + 2B \le 1000$ and the second is $2C + A \le 500$. Without these constraints, there is no limit to the volume of the box.

Figure 5.2 shows the Mathcad calculation that finds the maximum box volume. Note that the volume is large because it is expressed in mm^2. A cube with an edge length of 1000 mm has a volume of 10^9 mm^3.

There are two formatting issues to address before we continue. The first is that this problem requires maximizing volume rather than minimizing. The methods we have worked with so far have all assumed that the objective function is to be minimized. Indeed almost all optimization methods make this assumption. The solution is simple. Maximizing the objective function is the same as minimizing the negative of the objective function. Thus, maximizing an objective function $f(x_1, x_2, \ldots)$ is mathematically identical to minimizing $-f(x_1, x_2, \ldots)$.

© Springer International Publishing AG, part of Springer Nature 2018
M. French, *Fundamentals of Optimization*,
https://doi.org/10.1007/978-3-319-76192-3_5

Fig. 5.1 Layout of a box

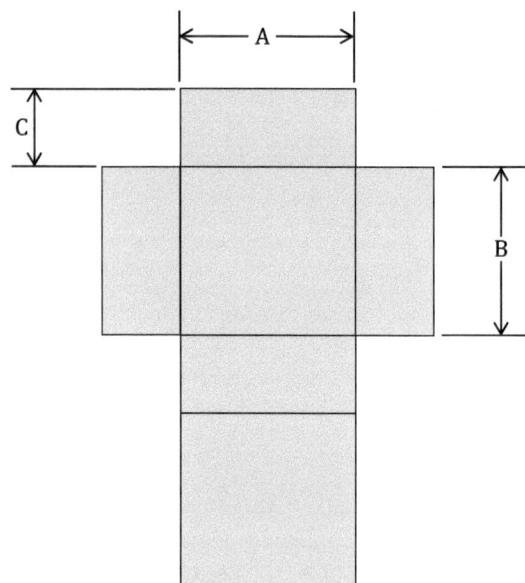

$$A := 1 \quad B := 1 \quad C := 1$$

Given

$$\text{Vol}(A, B, C) := A \cdot B \cdot C$$

$$2 \cdot C + 2 \cdot B \le 1000$$

$$2 \cdot C + A \le 500$$

$$\begin{pmatrix} A \\ B \\ C \end{pmatrix} := \text{Maximize}(\text{Vol}, A, B, C) = \begin{pmatrix} 288.675 \\ 394.338 \\ 105.663 \end{pmatrix} \quad \text{<= Box Dimensions}$$

$$\text{Vol}(A, B, C) = 1.203 \times 10^7 \quad \text{<= Maximum Volume}$$

Fig. 5.2 Mathcad calculation to maximize box volume

The second issue is how to write out constraints. The standard practice [1] is to write out each constraint as a function less than or equal to zero. There is a standard format for writing out optimization problems with inequality constraints. It looks like this:

$$\text{Minimize } F(x_1, x_2, \ldots, x_n)$$

Subject to
$$g_1 = g(x_1, x_2, \ldots, x_n) \le 0$$
$$g_2 = g(x_1, x_2, \ldots, x_n) \le 0$$
$$\vdots$$
$$g_m = g(x_1, x_2, \ldots, x_n) \le 0$$

A := 1   B := 1   C := 1

Given

$$Vol(A,B,C) := -A \cdot B \cdot C$$

$$2 \cdot C + 2 \cdot B - 1000 \leq 0$$

$$2 \cdot C + A - 500 \leq 0$$

$$\begin{pmatrix} A \\ B \\ C \end{pmatrix} := \text{Minimize}(\text{Vol}, A, B, C) = \begin{pmatrix} 288.675 \\ 394.338 \\ 105.663 \end{pmatrix} \quad \text{<= Box Dimensions}$$

$$Vol(A,B,C) = -1.203 \times 10^7 \quad \text{<= Maximum Volume}$$

Fig. 5.3 Box volume problem in standard form

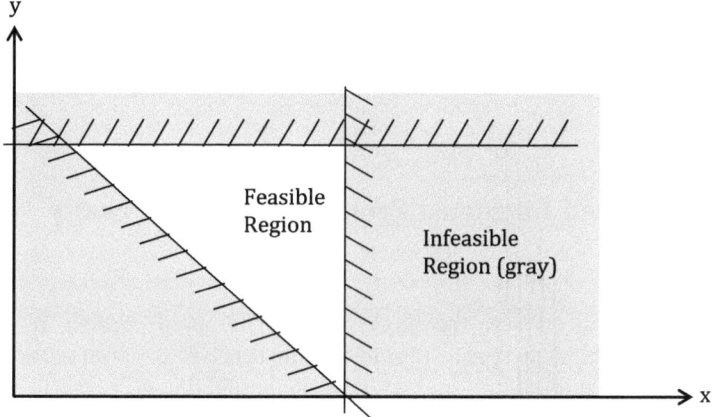

Fig. 5.4 Feasible and infeasible regions

The functions g_1 through g_m are the constraint functions. If $g \leq 0$, the constraint is said to be satisfied. If not, then the constraint is said to be violated. Figure 5.3 shows the same problem, but now in standard form.

The problem is mathematically unchanged but now uses the minimize function. The volume is negative as a result.

Adding constraints walls off portions of design space as shown in a two-variable example in Fig. 5.4. The area in which all the constraints are satisfied is called the feasible region. All the rest is called the infeasible region. By definition, the optimum lies inside the feasible region or at the boundary. In practical problems, optima often lie at the boundary of the feasible region.

It is routine to use software such as MATLAB or Mathcad to solve constrained optimization problems, but the inner workings are essentially hidden from the user. To see how optimization works in the presence of constraints, we'll need to understand how at least one constrained optimization method works. Perhaps the easiest constrained method to understand is the exterior penalty function method (EPF) [2]. There are certainly many constrained methods available and, while they use a wide range of underlying strategies, they can often look similar to the user.

Many engineering analysis methods transform a problem that is difficult to solve into a mathematically equivalent problem that is easier to solve. In a similar way, EPF transforms a constrained optimization problem into a nearly equivalent unconstrained problem. We already have tools for solving unconstrained problems and these can be used for the transformed constrained problem.

EPF modifies the objective function, adding a penalty for moving outside the feasible region. Minimizing this modified objective function naturally pulls the solution back into the feasible region. This modified function is called the pseudo objective function.

Note that EPF simply defines the pseudo objective function so that any unconstrained optimization method (like the ones in previous chapter) can be used to solve constrained problems. For example, EPF combined with steepest descent can be a robust way to solve constrained optimization problems.

5.2 Constrained Lifeguard Problem: Exterior Penalty Function

Let's make a modification to the single variable lifeguard problem by adding a constraint. Let's assume there is an obstacle on the beach at $x = 50$m so the lifeguard can't run beyond that point (my students have suggested that the obstacle might be a dead whale, yuck). Our knowledge of the physics of the problem might strongly suggest that the constrained optimum will be at $x^* = 50$ m, but let's assume for now that we don't know that.

Let's start by looking at design space in Fig. 5.5. The feasible region is white and the infeasible region is gray. Note that the boundary between the feasible and infeasible regions is the line defined by $g(x) = 0$.

The pseudo objective function is just the sum of the objective function and a penalty function. For this single variable, single-constraint problem, the pseudo objective function is

$$P(x) = F(x) + p(x) \tag{5.1}$$

where $F(x)$ is objective function from Eq. (2.3) and $p(x)$ is the penalty function

$$p(x) = R\Phi(g(x))[g(x)]^2 \tag{5.2}$$

R is a constant assigned by the user and Φ is a mathematical on-off switch called the Heaviside step function. Before going further, let's pause to look at the Heaviside

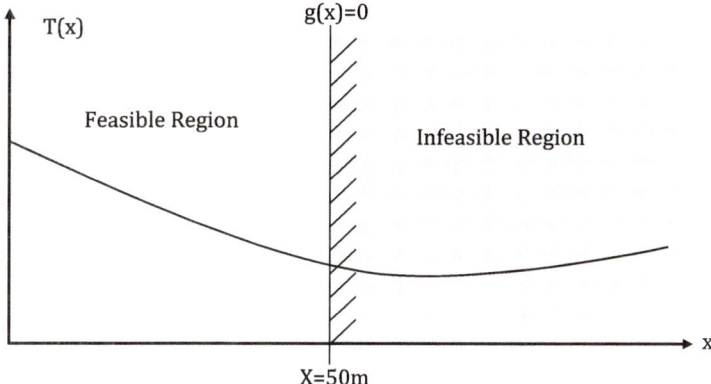

Fig. 5.5 Design space for constrained lifeguard problem

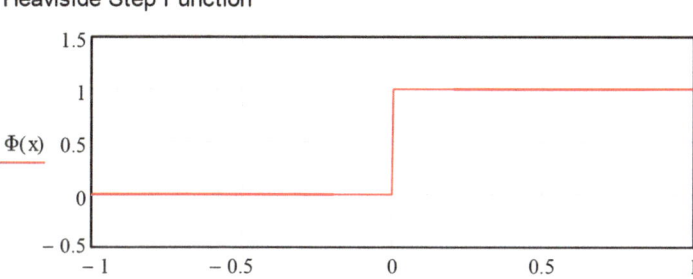

Fig. 5.6 Heaviside step function, $\Phi(x)$

step function. It was named after its originator, Oliver Heaviside, a self-taught electrical engineer, mathematician, and physicist who worked in England in the late 1800s and early 1900s [3].

The step function $\Phi(x)$ has a value of 0 when $x < 0$ and 1 when $x \geq 0$ as shown in Fig. 5.6. It acts as a switch that turns on when $x = 0$. By making $g(x)$ the argument of the step function, it becomes a switch that turns on when the constraint value is zero – at the constraint boundary. Thus, $\Phi = 1$ when the constraint is violated and the penalty function "turns on."

Because of the step function, the penalty function is zero when the constraint is satisfied and only adds to the objective function when the constraint is violated. So, when searching inside the feasible region, away from constraint boundaries, the objective function is unconstrained.

Figure 5.7 shows the objective function and the pseudo objective for the constrained lifeguard problem. Note how the penalty increases as R increases.

Since the penalty function doesn't turn on until the constraint is violated, the minimum value of the pseudo objective function will lie slightly outside the feasible

$$T(x) := \frac{1}{7} \cdot \sqrt{50^2 + x^2} + \frac{1}{2} \cdot \sqrt{50^2 + (100 - x)^2} \qquad g(x) := x - 50$$

$$P(R, x) := T(x) + R \cdot \Phi(g(x)) \cdot g(x)^2$$

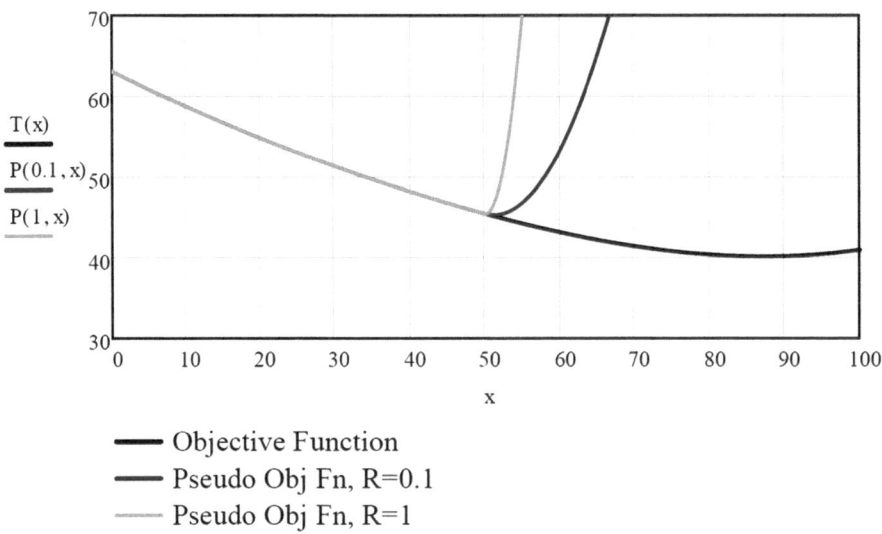

Fig. 5.7 Constrained lifeguard problem pseudo objective function

region. Figure 5.8 shows a close-up of the pseudo objective function in the region near $x = 50$ m.

The fact that EPF gives a solution slightly outside the feasible region is not necessarily a problem. Figure 5.8 shows that higher values of R push the solution closer to the boundary of the feasible region. Table 5.1 shows the effect R has on x^*. When R gets large enough, the solution is essentially on the constraint boundary.

Finally, the Heaviside step function is a mathematical ideal. It takes only one of two values (0 and 1) and has a derivative of zero everywhere except $x = 0$. At this point, the function is discontinuous. If this discontinuity causes numerical problems, the step function can be approximated by a number of continuous functions. One of the most familiar is the sigmoid function known as the Logistic function [4].

$$s(x) = \frac{1}{1 + e^{-cx}} \tag{5.3}$$

where c is a parameter that describes the slope at $x = 0$. Figure 5.9 shows the logistic function and the effect of increasing c.

$$T(x) := \frac{1}{7} \cdot \sqrt{50^2 + x^2} + \frac{1}{2} \cdot \sqrt{50^2 + (100 - x)^2} \qquad g(x) := x - 50$$

$$P(R, x) := T(x) + R \cdot \Phi(g(x)) \cdot g(x)^2$$

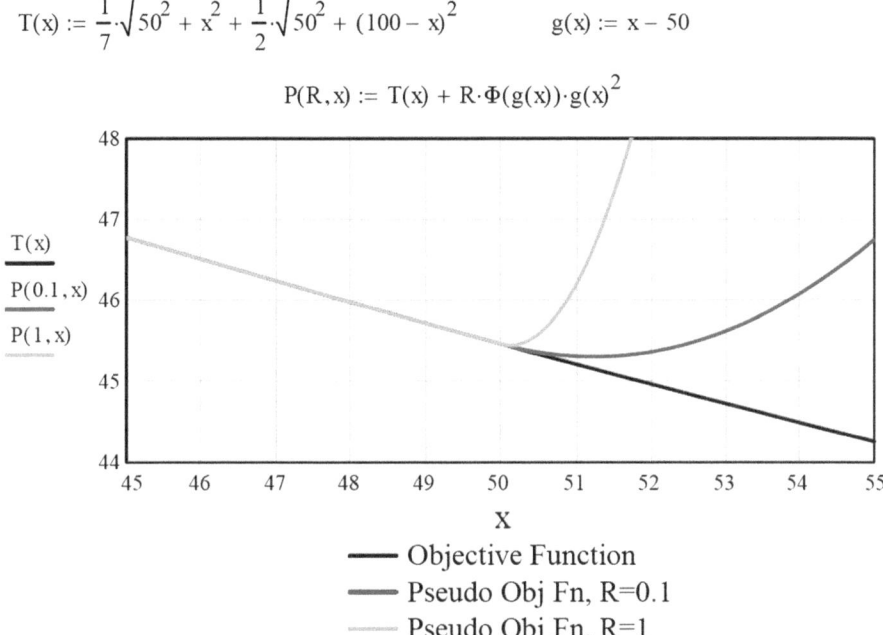

Fig. 5.8   Pseudo objective function near constraint boundary

Table 5.1   Effect of increasing R on x*

Penalty Factor, R	Optimum Point, x*
0.1	51.234
1.0	50.126
10	50.013
100	50.001

5.3   Minimum Surface Area of a Can

Drink cans are one of the most optimized products many of us use. Let's do a simple version of a can design problem and find the dimensions of a can that holds at least 500 mL (500 cm³) with the smallest possible surface area. The dimensions of the can are shown in Fig. 5.10.

The total surface area of the can is the area of the curved side plus the areas of the top and bottom.

$$A = \pi DH + 2\left(\frac{\pi}{4}D^2\right) = \pi DH + \frac{\pi}{2}D^2 \qquad (5.4)$$

and the volume of the can is

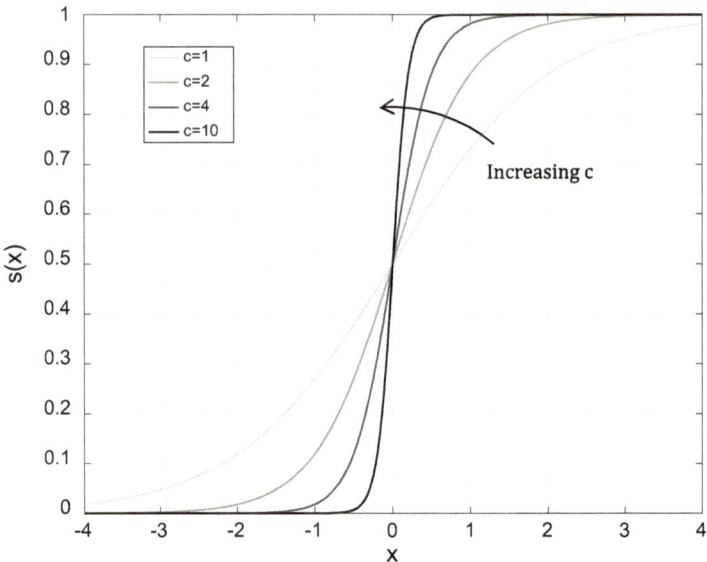

Fig. 5.9 Logistic function as an approximation to the Heaviside step function

Fig. 5.10 Dimensions of a can

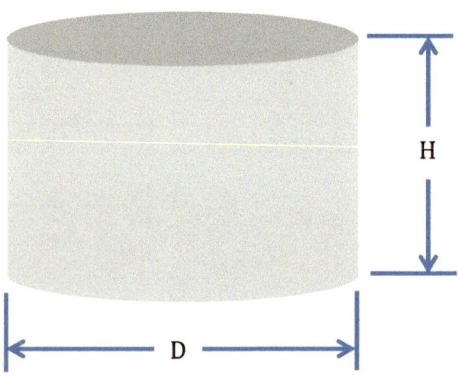

$$V = \frac{\pi}{4} D^2 H \tag{5.5}$$

The first step is to write out the formal optimization problem

$$\text{Minimize} : A(D, H) = \pi D H + \frac{\pi}{2} D^{2.}$$

$$\text{Subject to} : g(D, H) = -\frac{\pi}{4} D^2 H + 500 \leq 0$$

The general form of the pseudo objective function is

$$A(D,H) := \pi \cdot D \cdot H + \frac{\pi}{2} \cdot D^2 \qquad g(D,H) := \frac{-\pi}{4} \cdot D^2 \cdot H + 500 \qquad R := 1$$

$$P(D,H) := A(D,H) + R \cdot \Phi(g(D,H)) \cdot g(D,H)^2$$

$D := 500 \qquad H := 500 \qquad$ <= Start inside the feasible region

$$\begin{pmatrix} D_{star} \\ H_{star} \end{pmatrix} := \text{minimize}(P,D,H) = \begin{pmatrix} -2.684 \times 10^{10} \\ 2.129 \times 10^{10} \end{pmatrix}$$

$$A(D_{star}, H_{star}) = -6.635 \times 10^{20}$$

Fig. 5.11 Initial Mathcad solution of can area problem

$$P(D,H) = A(D,H) + R\Phi[g(D,H)][g(D,H)]^2 \qquad (5.6)$$

Let's start by letting Mathcad do the optimization and simply find an optimum can design as shown in Fig. 5.11.

Mathematically the solution is correct, but in physical terms, it is nonsense. The problem is that there is nothing in the problem statement that prevents the can from having a negative diameter, so the solver let it go negative. Let's add an additional constraint that requires $D > 0$.

$$\text{Minimize} : A(D,H) = \pi DH + \frac{\pi}{2} D^2$$

$$\text{Subject to} : g_1(D,H) = -\frac{\pi}{4} D^2 H + 500 \le 0$$

$$g_2(D) = -D \le 0$$

Figure 5.12 shows what happens when the additional constraint is added. The answer then makes both mathematical and physical sense. In practice, the solution found by Mathcad was slightly dependent on the choice of starting point. This is because the curvature of the objective function near the minimum is very low in one direction and a range of locations in deign space satisfied exit criteria.

Figure 5.13 shows design space for the can problem when $R = 1$. It is clear where the penalty function turns on. It is also clear that the objective function changes very little as we move along the constraint boundary. Thus, there is a range of $D*$ and $H*$ that would give a surface area close to the optimum.

This may be good news for people designing drink cans. It means that a relatively wide range of can aspect ratios will give surface areas near the minimum. Strictly speaking, the minimum area occurs when the height equals the diameter. However, such a can could be difficult to hold with one hand. A better solution might be a can

$$A(D,H) := \pi \cdot D \cdot H + \frac{\pi}{2} \cdot D^2 \qquad\qquad Vol(D,H) := \frac{\pi}{4} \cdot D^2 H \qquad\qquad R := 1$$

$$g_1(D,H) := \frac{-\pi}{4} \cdot D^2 \cdot H + 500 \qquad\qquad g_2(D) := -D$$

$$P(D,H) := A(D,H) + R \cdot \Phi\big(g_1(D,H)\big) \cdot g_1(D,H)^2 + R \cdot \Phi\big(g_2(D)\big) \cdot g_2(D)^2$$

$D := 10 \qquad\qquad H := 10 \qquad\qquad$ **<= Start inside the feasible region**

$$\begin{pmatrix} D_{star} \\ H_{star} \end{pmatrix} := minimize(P,D,H) = \begin{pmatrix} 8.557 \\ 8.691 \end{pmatrix}$$

$$A\big(D_{star},H_{star}\big) = 348.656 \qquad\qquad Vol\big(D_{star},H_{star}\big) = 499.812$$

Fig. 5.12 Minimum can area calculation with positive diameter constraint

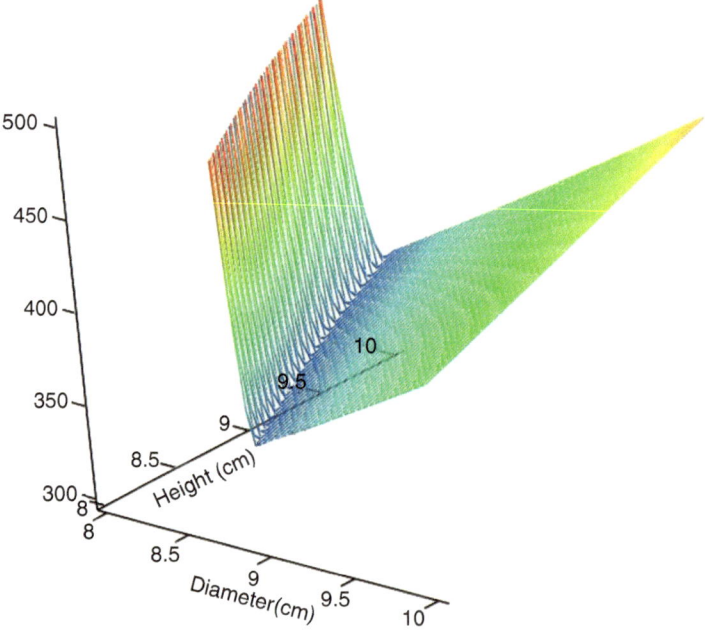

Fig. 5.13 Design space for can area problem, $R = 1$

with a diameter of 7.10 cm and a height of 12.6 cm. The resulting surface area is about 3.5% higher than the optimum, but the can would be much easier to hold, particularly for people with smaller hands or limited dexterity.

$$D := 15 \qquad H := 15 \qquad \text{<= Initial Guess}$$

$$A(D,H) := \pi \cdot D \cdot H + \frac{\pi}{2} \cdot D^2 \qquad \text{<= Define Objective Function}$$

Given

$$\frac{\pi}{4} \cdot D^2 \cdot H \geq 500 \qquad \text{<= Constraint Inside Solve Block}$$

$$\text{Minimize}(A,D,H) = \begin{pmatrix} 8.604 \\ 8.599 \end{pmatrix}$$

Fig. 5.14 Can problem with two constraints solved directly in Mathcad

Note that it is possible to define a constrained optimization problem in Mathcad directly, without having to use a pseudo objective function. Figure 5.14 shows a Mathcad solve block with the objective function and both constraints.

5.4   Equality Constraints

Not all constraints are inequality constraints. It is possible to have equality constraints of the form $g(x_1, x_2, \ldots, x_n) = c$ where c is a constant. In mathematical terms, this means that the minimum value of the objective lies along a single curve in design space. For example, if you suspect beforehand that the optimum lies at a specific constraint boundary, it is possible to turn that constraint into an equality constraint and only look there.

Any deviation from the equality constraint constitutes a violation, so the Heaviside step function isn't needed. Here is the minimum can area problem written out with one equality constraint and one inequality constraint

$$\text{Minimize} : A(D,H) = \pi DH + \frac{\pi}{2} D^2$$

$$\text{Subject to} : g_1(D,H) = -\frac{\pi}{4} D^2 H + 500 = 0$$

$$g_2(D) = -D \leq 0$$

The pseudo objective function also becomes simpler as only the second constraint needs the Heaviside step function.

$$P(D,H) = A(D,H) + R[g_1(D,H)]^2 + R\Phi[g_2(D)][g_2(D)]^2 \qquad (5.7)$$

In practice, equality constraints should be used sparingly as it can be easy to specify an over-constrained problem – one for which there is no solution. For example, using two equality constraints means the solution must lie on the point where the two curves cross one another. This is fine as long as the curves actually cross. If they do not, the problem is over-constrained and has no solution.

5.5   Approximate Solutions

A variation on curve fitting is finding approximate solutions to problems that are difficult to solve exactly. In this section, we'll explore one problem in which simple functions are used to find an approximate solution to a minimization problem. Then we'll see how simple assumed functions can be used to approximately solve a differential equation.

5.5.1   Hanging Chain Problem

In an earlier section, we saw that a hanging chain takes the form of a catenary [5]. The expression for a catenary is

$$c(x) = a\left(e^{x/a} + e^{-x/a}\right) + d \tag{5.8}$$

Let's consider a chain 4 m long hanging from two points that are separated by 2 m as shown in Fig. 5.15. For this chain, the coefficients are $a = -0.45928$ and $d = 2.05206$.

Fig. 5.15 Chain hanging from two points

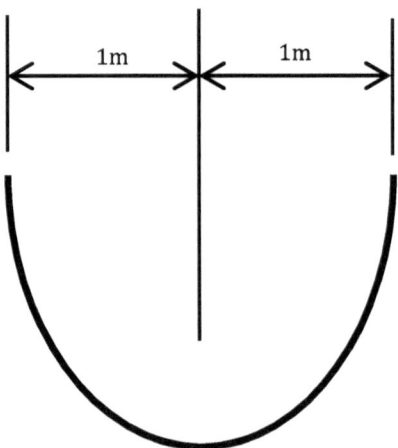

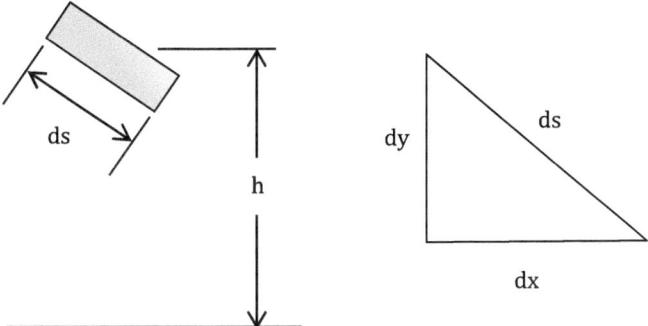

Fig. 5.16 Geometry of a chain segment

The catenary is the one shape out of all the possibilities that minimizes the potential energy of the chain. For simplicity, let's assume mg = 1; the solution is the same, no matter what the mass of the individual links is, so there is no loss of generality. In this case, the potential energy for the catenary is −3.64483. The hanging chain problem was one of the early successes in the development of the calculus of variations, a branch of mathematics devoted to finding minima and maxima. The details are well beyond the scope of this book. However, optimization can be used to find an approximate solution.

To start, we need an expression for the potential energy of the chain. We will do this by adding up the potential energy of segments of the chain. The geometry of a segment of the chain is shown in Fig. 5.16.

The Pythagorean theorem says that

$$ds^2 = dx^2 + dy^2 \tag{5.9}$$

Divide both sides by dx^2 to get

$$\frac{ds^2}{dx^2} = \frac{dx^2}{dx^2} + \frac{dy^2}{dx^2} \tag{5.10}$$

Now, multiply through by dx^2 to get

$$ds^2 = \left[\frac{dx^2}{dx^2} + \frac{dy^2}{dx^2}\right] dx^2 \tag{5.11}$$

Finally, take the square root of both sides to get

$$ds = \sqrt{\frac{dx^2}{dx^2} + \frac{dy^2}{dx^2}} dx \tag{5.12}$$

so the expression becomes

$$ds = \sqrt{1 + \left(\frac{dy}{dx}\right)^2}\, dx \tag{5.13}$$

Now, all we need to do is to add up the potential energies of all the segments. If the segments are small enough, the summation becomes an integral.

$$\text{PE} = \sum_{i=1}^{N} \text{mgh}_i ds \Rightarrow mg \int_{-1}^{1} y(x)\sqrt{1 + \left(\frac{dy}{dx}\right)^2}\, dx \tag{5.14}$$

So far, $y(x)$ can be any function. However, we need a function that has the following properties:

- Function must pass through the points $(-1, 0)$ and $(1, 0)$.
- The problem is symmetric about the Y axis, so the approximating function should be as well.
- The length of the curve between the endpoints must be 4.

Perhaps the simplest possible approximating function is a parabola, $y(x) = c_1 x^2 + c_0$. This function is already symmetric about the Y axis so that requirement is met. In order for $y(-1) = y(1) = 0$, $c_1 = -c_0$. This means that the approximating function becomes $y(x) = c_1(x^2 - 1)$. The final requirement is that the length of the curve is 4. Enforcing this requirement gives

$$y_1 = 1.6346\left(x^2 - 1\right) \tag{5.15}$$

The obvious problem is that there is no design variable that can be changed to vary the potential energy – this function has no design variables. However, it is possible to find two approximating functions like this and then use optimization to find the proportions of them that minimize potential energy.

Another approximating function is an ellipse. Like the parabola, it is symmetric. Satisfying the requirements as before gives a second possible function.

$$y_2 = -1.51962\sqrt{1 - x^2} \tag{5.16}$$

Like the parabola, this function has no variable parameters that would change its shape.

A solution to the problem of identifying design variables is to sum the parabola and the ellipse with weighting factors. The final approximating function is then $F(x) = ay_1 + by_2$. The Mathcad solution is shown in Fig. 5.17. The exact solution superimposed over the approximate solution is shown in Fig. 5.18. To improve the approximate solution further, you could add more terms to the approximating function.

$A_1 := 1 \qquad A_2 := 1 \qquad y_1(x) \to 1.6346 \cdot x^2 - 1.6346 \qquad y_2(x) \to -1.51962 \cdot \sqrt{1 - x^2}$

$f(x, A_1, A_2) := A_1 \cdot y_1(x) + A_2 \cdot y_2(x) \to A_1 \cdot \left(1.6346 \cdot x^2 - 1.6346\right) + -1.51962 \cdot A_2 \cdot \sqrt{1 - x^2}$

Given

$$PE(A_1, A_2) := \int_{-1}^{1} f(x, A_1, A_2) \cdot \sqrt{1 + \left(\frac{d}{dx} f(x, A_1, A_2)\right)^2} \, dx \qquad \int_{-1}^{1} \sqrt{1 + \left(\frac{d}{dx} f(x, A_1, A_2)\right)^2} \, dx = 4$$

$$\binom{A_1}{A_2} := \text{Minimize}(PE, A_1, A_2) = \binom{0.82494}{0.1793}$$

$PE(A_1, A_2) = -3.63971 \qquad$ <= Potential energy of approximate solution

$\dfrac{PE(A_1, A_2)}{PE_{exact}} = 0.99859 \qquad$ <= Very close to minimum potential energy

Fig. 5.17 Mathcad solution to hanging chain approximation

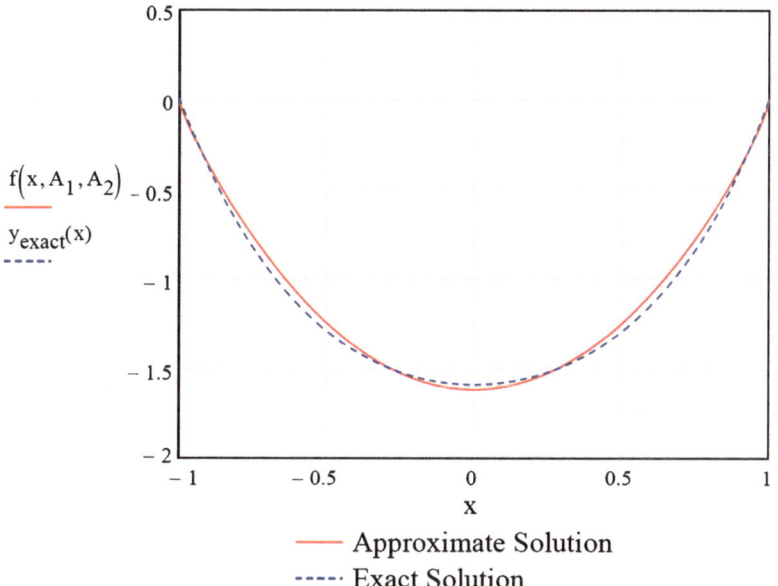

Fig. 5.18 Exact and approximate solutions to hanging chain problem

References

1. Luenberger DG, Ye Y (2008) Linear and nonlinear programming. Springer, Cham
2. Betts JT (1975) An improved penalty function method for solving constrained parameter optimization problems. J Optim Theory Appl 16(1/2):1–24
3. Nahin PJ (2002) Oliver Heaviside: the life, work and times of and electrical genius of the Victorian age. Johns Hopkins University Press, Baltimore
4. Richards FJ (1959) A flexible growth function for empirical use. J Exp Biol 10(29):290–300
5. Osserman R (2010) Mathematics of the gateway arch. Not Am Math Soc 57(2):220–229

Chapter 6
General Conditions for Solving Optimization Problems: Karush-Kuhn-Tucker Conditions

Each problem that I solved became a rule, which served afterwards to solve other problems

-Rene DesCartes

As you read more about optimization, you will undoubtedly run across the Kuhn-Tucker conditions, also called the Karush-Kuhn-Tucker (KKT) conditions [1]. The change is because scholars later discovered Karush's work and that it predated the publication by Kuhn and Tucker [2].

The KKT conditions are the formal mathematical conditions that define a maximum or minimum of an objective function. They also account for both equality and inequality constraints. If the problem at hand is written in a closed form, the KKT conditions can be used to solve the problem directly – no need to search through design space.

6.1 General Form of KKT Conditions

It shouldn't be surprising that there is an entire body of literature defining the mathematical underpinnings of optimization. While a complete survey is far too much for this book, it is helpful to learn something about the KKT conditions since they form the basis of many optimization algorithms. In cases where the problem can be written out in a closed form, it can be possible to find the optimum design by directly solving the KKT equations.

Here's the most general form of an optimization problem

$$
\begin{array}{lll}
\text{minimize} & f(x) & x = x_1, x_2, \cdots, x_n \\
\text{subject to} & g_i(x) \leq 0 & i = 1, \cdots, p \\
& h_j(x) = 0 & j = 1, \cdots, q
\end{array}
\tag{6.1}
$$

The objective function and the constraints can be combined to form a Lagrange function

© Springer International Publishing AG, part of Springer Nature 2018
M. French, *Fundamentals of Optimization*,
https://doi.org/10.1007/978-3-319-76192-3_6

$$L(x, \lambda, \mu) = f(x) + \sum_{i=1}^{p} \lambda_i g_i(x) + \sum_{j=1}^{q} \mu_j h_j(x) \tag{6.2}$$

where λ and μ are constants called Lagrange multipliers. The Karush-Kuhn-Tucker optimality conditions are

$$
\begin{aligned}
\frac{\partial}{\partial x_k} L(x, \lambda, \mu) &= 0 & k &= 1, \cdots, n \\
\lambda_i &\geq 0 & i &= 1, \cdots, p \\
\lambda_i g_i(x) &= 0 & i &= 1, \cdots, p \\
g_i(x) &\leq 0 & i &= 1, \cdots, p \\
h_i(x) &= 0 & j &= 1, \cdots, q
\end{aligned}
\tag{6.3}
$$

Solving this set of equations gives the solution to the constrained optimization problem. Not every optimization problem lends itself to directly solving the equations defined by the KKT conditions. However, they define what a solution means, so it's important to understand what these equations say and how they can be applied.

6.2   Application to an Unconstrained Problem of One Variable

It's worth spending a little time figuring out what the KKT conditions are and how they apply to the range of possible optimization problems. Let's start with the simplest class of problems and then extend the basic ideas to cover more sophisticated ones.

The simplest possible optimization problem is finding the minimum of an unconstrained function of one variable. Consider the function

$$y(x) = \sqrt{(x-2)^2 + \sin^2(x)} \tag{6.4}$$

Figure 6.1 shows that it is a simple nonlinear function with a clear minimum. At x^*, $df/dx = 0$, so the first element of the KKT conditions in Equation 6.3 is satisfied. The remaining elements don't apply since there are no constraints.

As all students learn in introductory calculus, two conditions must be true at the minimum of a function: the slope must be zero and the curvature must be positive. It is worth pausing for a moment to revisit the ideas of slope and curvature.

Students typically have few problems with the ideas of positive and negative slope. It is usually intuitive enough that positive slope means going uphill when moving in the positive x direction. Conversely, negative slope means going downhill.

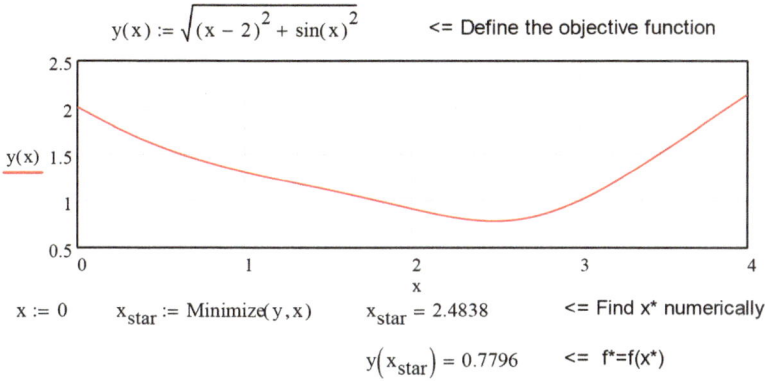

$$y(x) := \sqrt{(x-2)^2 + \sin(x)^2} \qquad \text{<= Define the objective function}$$

$x := 0 \qquad x_{star} := \text{Minimize}(y,x) \qquad x_{star} = 2.4838 \qquad \text{<= Find x* numerically}$

$$y(x_{star}) = 0.7796 \qquad \text{<= f*=f(x*)}$$

Fig. 6.1 Nonlinear function of one variable

Fig. 6.2 Conventions for negative and positive curvature

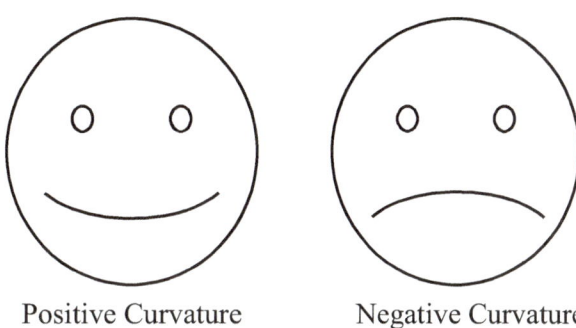

Positive Curvature   Negative Curvature

Curvature is sometimes more of a problem in that the ideas of positive and negative curvature are less intuitive. Figure 6.2 is perhaps a little silly, but students, having seen it, seldom forget it.

Thus, the necessary conditions for a minimum are $f'(x) = 0$ and $f''(x) > 0$. Figure 6.3 shows the calculation using Mathcad.

The next step in defining the necessary conditions for a minimum is to extend the ideas of slope and curvature to problems of more than one variable.

6.3 Application to an Unconstrained Problem of Two Variables

The optimality conditions for a function of one variable are easy to generalize to a function of several variables. Most mathematical concepts are valid, though sometimes with small modifications, to any number of dimensions and such is the case with slope and curvature. Let's start with a function of two variables

Find minimum by solving for point at which slope is zero

$$dydx(x) := \frac{d}{dx}y(x) \rightarrow \frac{2 \cdot x + 2 \cdot \cos(x) \cdot \sin(x) - 4}{2 \cdot \sqrt{\sin(x)^2 + (x-2)^2}}$$

$x_{min} := \text{root}(dydx(x), x) = 2.4838$ <= Result matches numerical solution

Confirm this is a minimum by ensuring that curvature is positive

$$d2ydx2(x) := \frac{d^2}{dx^2}y(x) \rightarrow \frac{2 \cdot \cos(x)^2 - 2 \cdot \sin(x)^2 + 2}{2 \cdot \sqrt{\sin(x)^2 + (x-2)^2}} - \frac{(2 \cdot x + 2 \cdot \cos(x) \cdot \sin(x) - 4)^2}{4 \cdot \left[\sin(x)^2 + (x-2)^2\right]^{\frac{3}{2}}}$$

$d2ydx2(x_{min}) = 1.6065$ <= Curvature is positive, since f''(x*) >0

Fig. 6.3 Slope and curvature of test function at minimum

$$f(x, y) = \frac{x - y}{(x^2 + 5)(y^2 + 5)} + \frac{y^2}{20000} \tag{6.5}$$

This is the butterfly function, shown in Appendix B. As is clear in Fig. 6.4, the minimum is near $(-1.3, 1.3)$. It gets its name because my students thought the contour plot looked like a butterfly. Who am I to disagree?

The first necessary condition for a minimum point is that the gradient – the multidimensional analog to the slope – is zero. Again, this is the first of the KKT conditions. For a single variable, the second necessary condition was that the curvature must be positive. In more than one dimension, the analogous requirement is that the matrix of second derivatives – called the Hessian – is positive definite. A positive definite matrix is one whose eigenvalues are positive. For two variables, the Hessian is

$$H(x, y) = \begin{bmatrix} \dfrac{\partial^2 f}{\partial x^2} & \dfrac{\partial^2 f}{\partial x \partial y} \\ \dfrac{\partial^2 f}{\partial x \partial y} & \dfrac{\partial^2 f}{\partial y^2} \end{bmatrix} \tag{6.6}$$

The extension to more design variables just continues the pattern established here. Note that the Hessian is symmetric, no matter what the objective function is. Figure 6.5 shows the calculation of both necessary conditions for the minimum.

The necessary conditions shown here are direct extensions of those for a single variable function. However, they don't account for the presence of constraints. The next extensions generalize the concept of optimality conditions by including constraints.

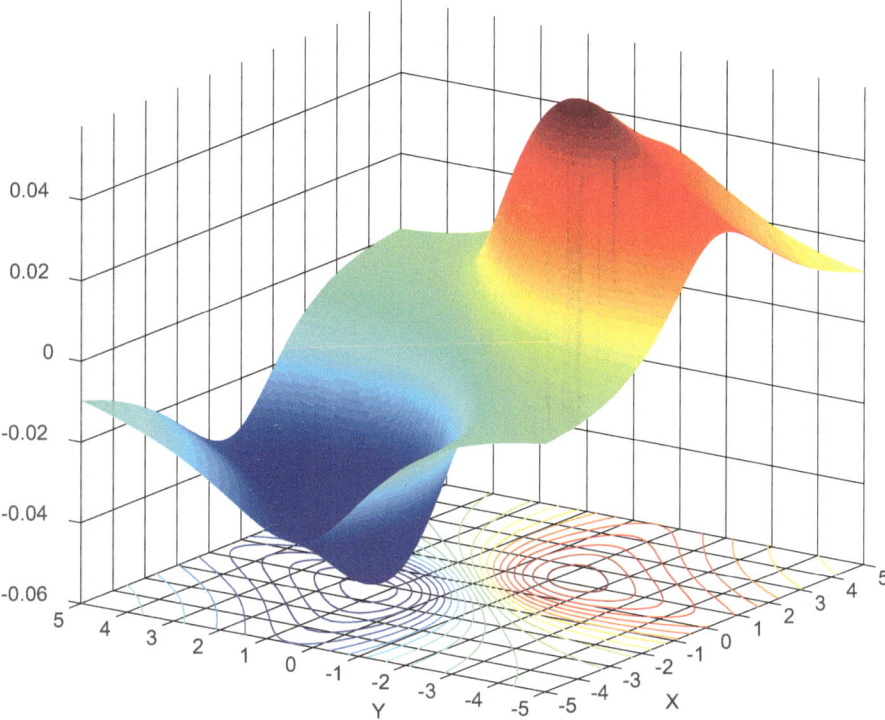

Fig. 6.4 Butterfly function showing both surface and contour plots

6.4 Application to a Single Variable Problem with an Equality Constraint

Having shown the optimality conditions for the unconstrained problem, let's modify the butterfly function problem by adding the equality constraint $y = -\sin(x)$. The formal optimization statement is then

$$\text{Minimize} \quad f(x,y) = \frac{x-y}{(x^2+5)(y^2+5)} + \frac{y^2}{20000} \tag{6.7}$$
$$\text{Subject to} \quad g(x) = y + \sin(x) = 0$$

Because this is an equality constraint, the minimum must lie on the curve $g(x)$. Figure 6.6 shows the constraint superimposed over the contour plot of the objective function. The constrained minimum is shown by a "+."

Of course, a straightforward way to solve the problem is to make the substitution $y = -\sin(x)$ directly into the objective function, reducing it to a function of x only.

Define Butterfly Function: $\quad f(x,y) := \dfrac{x - y}{\left(x^2 + 5\right)\cdot\left(y^2 + 5\right)} + \dfrac{y^2}{20000}$

Define Gradient:

$$g(x,y) := \nabla_{x,y} f(x,y) \rightarrow \begin{bmatrix} \dfrac{1}{\left(x^2+5\right)\cdot\left(y^2+5\right)} - \dfrac{2\cdot x\cdot(x-y)}{\left(x^2+5\right)^2\cdot\left(y^2+5\right)} \\[3mm] \dfrac{y}{10000} - \dfrac{1}{\left(x^2+5\right)\cdot\left(y^2+5\right)} - \dfrac{2\cdot y\cdot(x-y)}{\left(x^2+5\right)\cdot\left(y^2+5\right)^2} \end{bmatrix}$$

Define Hessian:

$$H(x,y) := \begin{bmatrix} \dfrac{d^2}{dx^2}f(x,y) & \dfrac{d}{dx}\!\left(\dfrac{d}{dy}f(x,y)\right) \\[3mm] \dfrac{d}{dx}\!\left(\dfrac{d}{dy}f(x,y)\right) & \dfrac{d^2}{dy^2}f(x,y) \end{bmatrix}$$

Solve for Minimum, grad(f)=0

$$x := -1 \quad y := 0 \qquad \text{Given} \qquad g(x,y) = \begin{pmatrix} 0 \\ 0 \end{pmatrix} \qquad \begin{pmatrix} x_{star} \\ y_{star} \end{pmatrix} := \text{find}(x,y) = \begin{pmatrix} -1.296 \\ 1.281 \end{pmatrix}$$

Verify that Hessian is Positive Definite - Eigenvalues are Positive

$$H\left(x_{star}, y_{star}\right) = \begin{pmatrix} 0.017 & -8.747 \times 10^{-3} \\ -8.747 \times 10^{-3} & 0.018 \end{pmatrix} \qquad \text{eigenvals}\left(H\left(x_{star}, y_{star}\right)\right) = \begin{pmatrix} 8.796 \times 10^{-3} \\ 0.026 \end{pmatrix}$$

Fig. 6.5 Calculation of necessary conditions for minimum of unconstrained function

$$f(x) = \frac{x + \sin(x)}{(x^2 + 5)(\sin^2(x) + 5)} + \frac{\sin^2(x)}{20000} \tag{6.8}$$

Figure 6.7 shows the resulting plot. The minimum lies at $x^* = -1.476$.

A more direct way to show the location of the minimum is to cut the surface plot of the objective function along the constraint boundary as shown in Fig. 6.8.

A mathematical description can start with a thought experiment. Imagine walking along the constraint curve in Fig. 6.6. As you walk along, you will cross contour lines until you reach a place where the contour lines are parallel to the constraint boundary. This is likely a constrained maximum or a constrained minimum, though some surfaces also have saddle points, where the gradient (slope) is zero but the curvature is not positive (Hessian is not positive definite).

If the constraint is parallel to the contour lines, then the gradient of the objective function must be proportional to the gradient of the constraint.

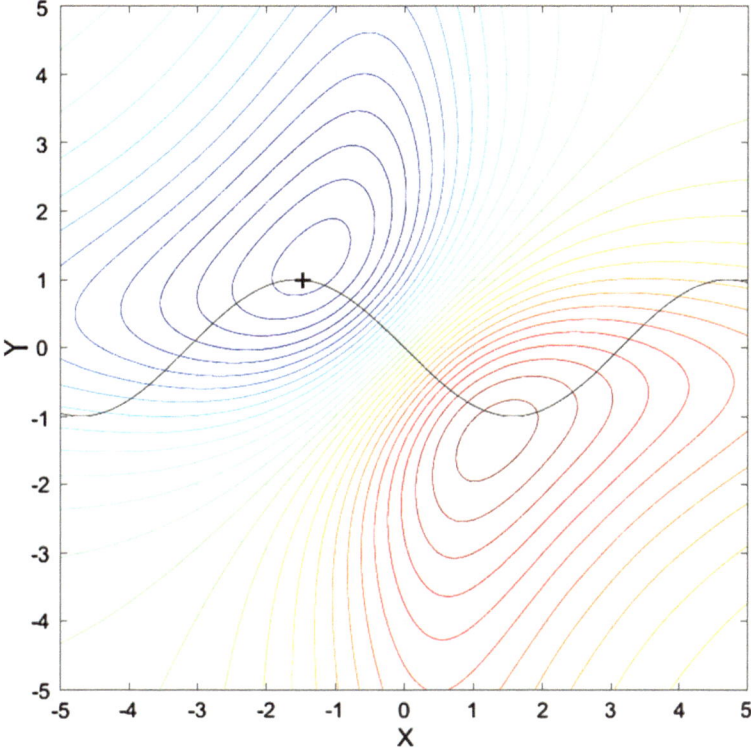

Fig. 6.6 Butterfly function with constraint superimposed

$$\nabla f = -\lambda \nabla g \qquad (6.9)$$

where λ is called a Lagrange multiplier. The negative sign is traditional and is sometimes omitted. The Lagrange multiplier is required because, although the directions of the two gradients are the same, their magnitudes are generally not. The additional constant just scales the constraint gradient so that it is equal to the objective function gradient. The result is often written as a Lagrange function or Lagrangian

$$L(x, y) = f(x, y) + \lambda g(x, y) \qquad (6.10)$$

The minimum or maximum of the Lagrangian is found in the same way as for the unconstrained function

$$\nabla L(x, y) = \nabla f(x, y) + \lambda g(x, y) = 0 \qquad (6.11)$$

Expanding this expression gives three equations, which allow solving for three variables, x, y, and λ.

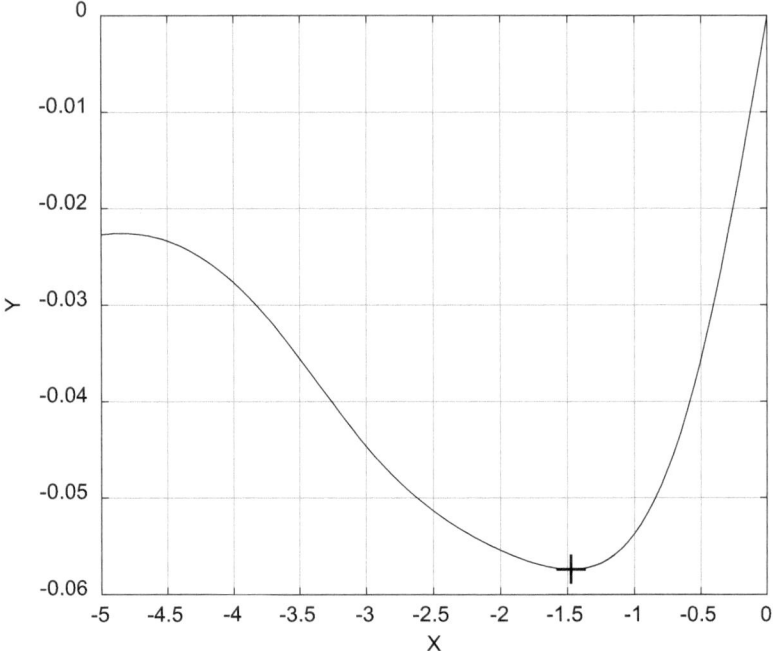

Fig. 6.7 Butterfly function with equality constraint substituted in $x^* = -1.476$

$$\frac{\partial L}{\partial x} = \frac{\partial f}{\partial x} + \lambda \frac{\partial g}{\partial x} = 0$$
$$\frac{\partial L}{\partial y} = \frac{\partial f}{\partial y} + \lambda \frac{\partial g}{\partial y} = 0 \qquad (6.12)$$
$$\frac{\partial L}{\partial \lambda} = g(x, y) = 0$$

The third of these equations enforces the equality constraint, making sure the solution lies on the constraint curve. Figure 6.9 shows the calculations to locate the constrained minimum using the method of Lagrange multipliers.

This method works for any number of variables and any number of constraints, though, in practice, more than a few equality constraints can easily create a problem for which there is no feasible region, so no constrained minimum can be found.

6.5   Application to a Multivariable Problem with Inequality Constraints

The last step in defining the optimality conditions is to extend them to inequality constraints of the form $g(x, y) \leq 0$. The Lagrangian in this case is identical to that for the equality constraint assuming that the constrained minimum still lies on the

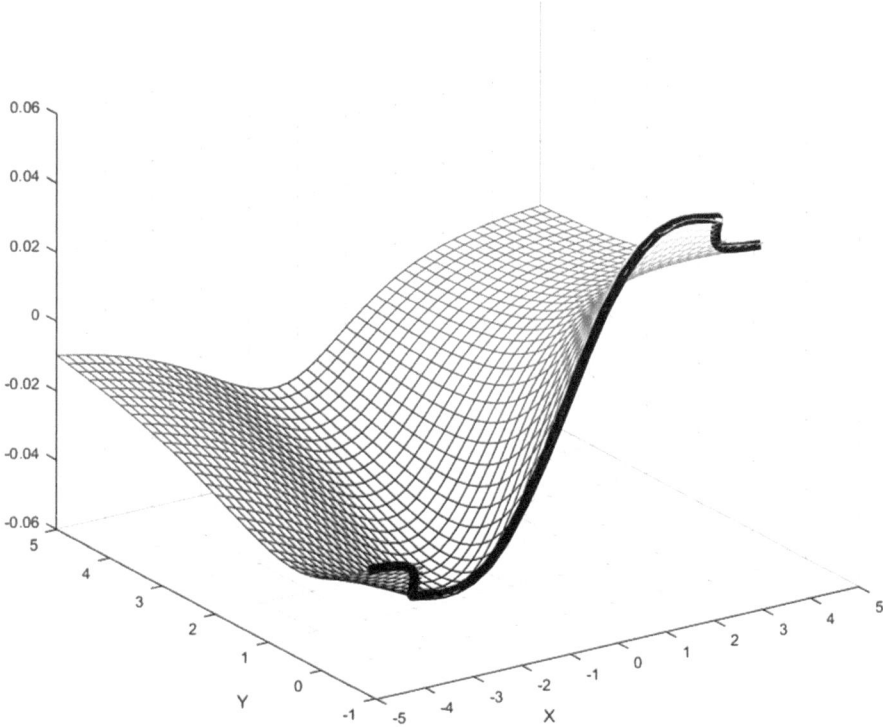

Fig. 6.8 Butterfly function with cutaway to show constraint boundary

constraint boundary. If the minimum value of the constrained problem lies at the boundary, then $g(x, y)$ is active and the inequality constraint gives the same solution as the equality constraint.

$$L(x, y) = f(x, y) + \lambda g(x, y) \qquad (6.13)$$

The other possibility is that the constraint is inactive and the problem is essentially unconstrained. The distinction is defined by the sign of the Lagrange multiplier, λ. If λ is positive, the constraint is active and should be retained as part of the Lagrangian. If λ is not positive, the constraint is not active and should be omitted from the Lagrangian.

One way to sort this out is to look at the problem graphically. At the constraint boundary, where $g(x) = 0$, the slope of the objective function and the slope of an active constraint must be opposite one another as shown in Fig. 6.10. At the constraint boundary, ∇g and ∇f have opposite signs, so the only way the expression $\nabla f + \lambda \nabla g = 0$ could be true is if $\lambda > 0$.

Define objective function and equality constraint

$$f(x,y) := \frac{x - y}{\left(x^2 + 5\right)\cdot\left(y^2 + 5\right)} + \frac{y^2}{20000} \qquad\qquad g(x,y) := y + \sin(x)$$

$$L(x,y,\lambda) := f(x,y) + \lambda\cdot g(x,y) \qquad \text{<= Define Lagrangian}$$

Define gradient of Lagrangian

$$\mathrm{GradL}(x,y,\lambda) := \nabla_{x,y,\lambda} L(x,y,\lambda) \rightarrow \begin{bmatrix} \dfrac{1}{\left(x^2 + 5\right)\cdot\left(y^2 + 5\right)} + \lambda\cdot\cos(x) - \dfrac{2\cdot x\cdot(x - y)}{\left(x^2 + 5\right)^2\cdot\left(y^2 + 5\right)} \\[3mm] \lambda + \dfrac{y}{10000} - \dfrac{1}{\left(x^2 + 5\right)\cdot\left(y^2 + 5\right)} - \dfrac{2\cdot y\cdot(x - y)}{\left(x^2 + 5\right)\cdot\left(y^2 + 5\right)^2} \\[3mm] y + \sin(x) \end{bmatrix}$$

$$x := 0 \qquad y := 0 \qquad \lambda := 0 \qquad \text{<= Initial guesses, required by solver}$$

Solve gradient equations to find location of minimum

$$\text{Given} \quad \mathrm{GradL}(x,y,\lambda) = \begin{pmatrix} 0 \\ 0 \\ 0 \end{pmatrix} \qquad \begin{pmatrix} x_{\mathrm{star}} \\ y_{\mathrm{star}} \\ \lambda_{\mathrm{star}} \end{pmatrix} := \mathrm{Find}(x,y,\lambda) = \begin{pmatrix} -1.476 \\ 0.996 \\ 4.052 \times 10^{-3} \end{pmatrix}$$

Fig. 6.9 Finding the constrained minimum using method of Lagrange multipliers

Let's revisit the previous example and replace the equality constraint with an inequality constraint. There are two possibilities for this case. The first is that the feasible region is the region below the constraint boundary as shown in Fig. 6.11. In this case, the constraint is defined as

$$g(x, y) = y + \sin(x) \leq 0 \tag{6.14}$$

The solution to the equation $\nabla L = 0$ is exactly the same as for the equality constraint (solution shown in Fig. 6.9) since it is essentially solving the same problem. Again, $\lambda = 4.052 \times 10^{-3}$, with the positive Lagrange multiplier telling us that the constraint is active.

A quick way to tell which side of the constraint boundary is the feasible region is to pick a point that is easy to evaluate. For example, the point $(0, 1)$ is easy to evaluate and clearly on one side of the boundary. $g(0, 1) = 1$, so the constraint is not satisfied and point $(0, 1)$ is not in the feasible region.

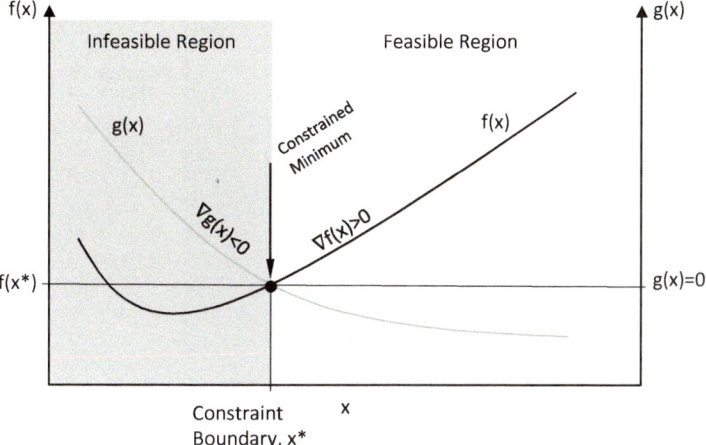

Fig. 6.10 Geometry of the constrained minimum

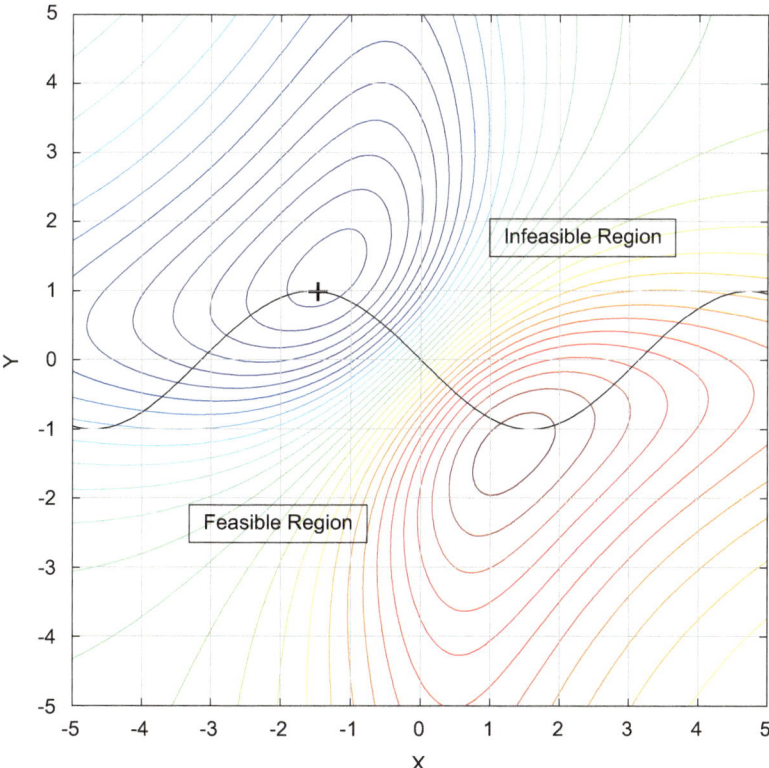

Fig. 6.11 Inequality constraint $g(x, y) = y + \sin(x) \leq 0$

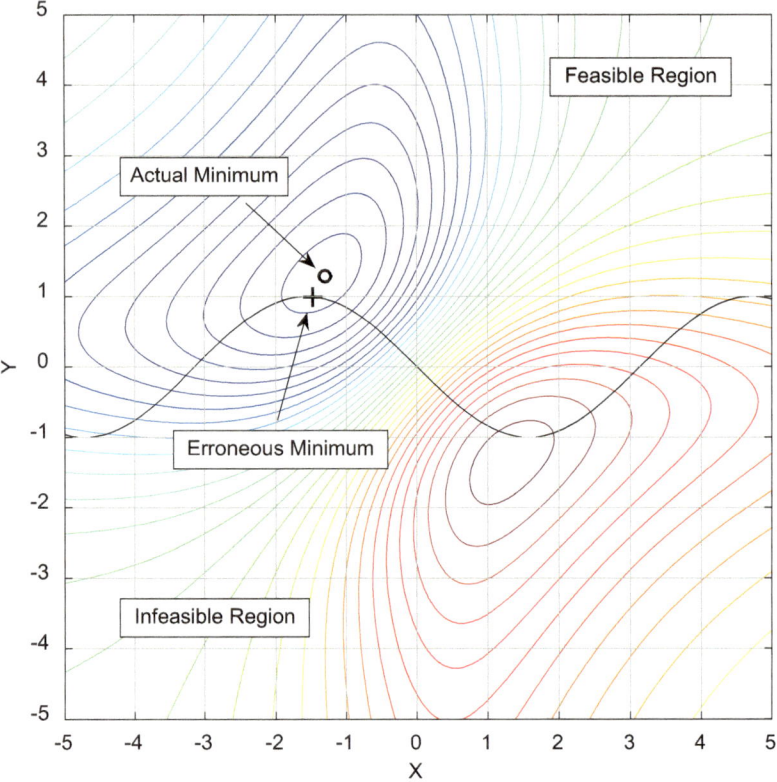

Fig. 6.12 Inequality constraint $g(x, y) = -y - \sin(x) \leq 0$

The other possible form of the constraint is $g(x, y) \geq 0$ which, written out in standard form, is

$$g(x, y) = -y - \sin(x) \leq 0 \qquad (6.15)$$

The feasible and infeasible regions are now reversed as shown in Fig. 6.12. Note that, without some modification, the predicted minimum point hasn't changed – it still lies on the constraint boundary – even though the global minimum clearly lies in the feasible region and away from the constraint boundary. This is because the constraint has been retained in the Lagrangian, even though it is not active.

Figure 6.13 shows the corresponding calculation of $\nabla L = 0$. While x^* and y^* haven't changed, the Lagrange multiplier has. It is now negative, indicating that the slopes of ∇f and ∇g are in the same direction. In more direct terms, it means this is

$$f(x,y) := \frac{x - y}{\left(x^2 + 5\right)\cdot\left(y^2 + 5\right)} + \frac{y^2}{20000} \qquad g(x,y) := -y - \sin(x)$$

$$L(x,y,\lambda) := f(x,y) + \lambda\cdot g(x,y) \qquad \text{<= Define Lagrangian}$$

Define gradient of Lagrangian

$$GradL(x,y,\lambda) := \nabla_{x,y,\lambda} L(x,y,\lambda) \rightarrow \begin{bmatrix} \dfrac{1}{\left(x^2 + 5\right)\cdot\left(y^2 + 5\right)} - \lambda\cdot\cos(x) - \dfrac{2\cdot x\cdot(x - y)}{\left(x^2 + 5\right)^2\cdot\left(y^2 + 5\right)} \\[2em] \dfrac{y}{10000} - \lambda - \dfrac{1}{\left(x^2 + 5\right)\cdot\left(y^2 + 5\right)} - \dfrac{2\cdot y\cdot(x - y)}{\left(x^2 + 5\right)\cdot\left(y^2 + 5\right)^2} \\[2em] -y - \sin(x) \end{bmatrix}$$

$$x := 0 \qquad y := 0 \qquad \lambda := 0 \qquad \text{<= Initial guesses, required by solver}$$

Solve gradient equations to find location of minimum

$$\text{Given} \quad GradL(x,y,\lambda) = \begin{pmatrix} 0 \\ 0 \\ 0 \end{pmatrix} \quad \begin{pmatrix} x_{star} \\ y_{star} \\ \lambda_{star} \end{pmatrix} := Find(x,y,\lambda) = \begin{pmatrix} -1.476 \\ 0.996 \\ -4.052 \times 10^{-3} \end{pmatrix}$$

Negative Lagrange multiplier shows that the constraint is not active

Fig. 6.13 Finding the erroneous constrained minimum where $g(x, y) = -y - \sin(x) \leq 0$

effectively an unconstrained problem since the constraint is not active at the minimum point (x^*, y^*).

We have looked at cases where $\lambda > 0$ and $\lambda < 0$, but it is also possible that $\lambda = 0$. Since $\nabla f = -\lambda \nabla g$, $\lambda = 0$ when $\nabla f = 0$. That is, the gradient of the objective function a x^* is zero in the presence of an equality constraint.

6.6 Multiple Constraints

Having multiple constraints is a routine situation and the conditions for optimality must reflect that. In general, terms can simply be added to the Lagrangian to account for additional constraints.

$$L = f + \lambda_1 g_1 + \lambda_2 g_2 + \lambda_3 g_3 + \cdots \tag{6.16}$$

As always, though, the possibility remains that adding too many constraints will result in a problem with no feasible region.

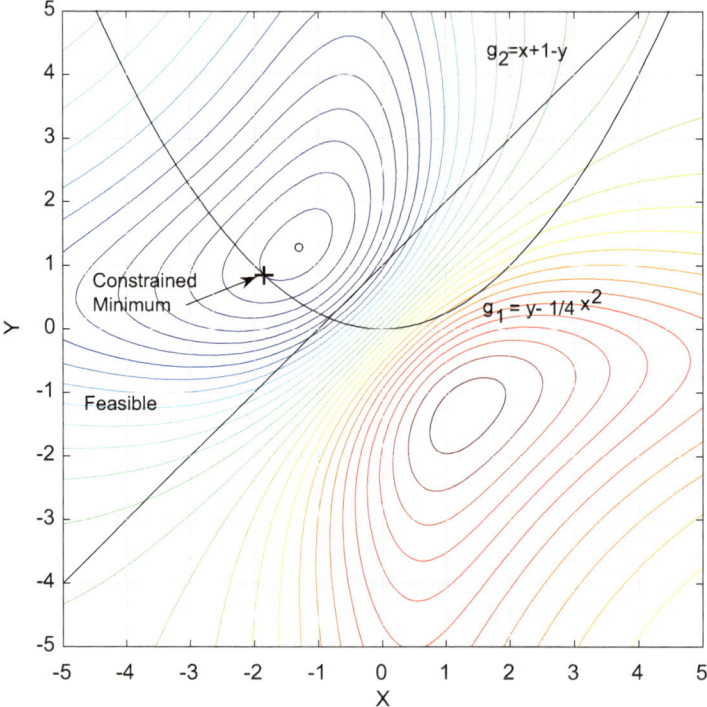

Fig. 6.14 Butterfly function with two inequality constraints

Let's stick with the butterfly function and consider the condition where there are two constraints. Let the two constraints be.

$$
\begin{aligned}
g_1(x, y) &= y - x^2 \leq 0 \\
g_2(x, y) &= x + 1 - y \leq 0
\end{aligned}
\tag{6.17}
$$

In this case, the location of the unconstrained minimum violates g_1 and satisfies g_2 as shown in Fig. 6.14. Because of the shape of this particular objective function, this means that the location of the constrained minimum will lie on the boundary of g_1. The second constraint, g_2, will not be active. To illustrate, it is simple to insert the new constraints into the previous calculations and again solve for $\nabla L = 0$, as shown in Fig. 6.15. Note that the Lagrange function containing both constraints erroneously gives a solution at the intersection of the constraints. Removing the inactive constraint, g_2, gives the correct constrained minimum.

It is certainly possible to define both equality and inequality constraints for the same problem. Doing that results in Eq. (6.2) above and gives the KKT conditions in Eq. (6.3). Finally, we should note that this is a brief introduction to the KKT conditions. Readers wishing more depth should explore the Lagrange dual function and the concept of the dual problem.

Butterfly function with two constraints

$$f(x,y) := \frac{x - y}{\left(x^2 + 5\right)\cdot\left(y^2 + 5\right)} + \frac{y^2}{20000} \qquad g_1(x,y) := y - \frac{1}{4}\cdot x^2 \qquad g_2(x,y) := x + 1 - y$$

$$L\left(x,y,\lambda_1,\lambda_2\right) := f(x,y) + \lambda_1\cdot g_1(x,y) + \lambda_2\cdot g_2(x,y) \qquad \Leftarrow \text{ Define Lagrangian}$$

Define gradient of Lagrangian

$$\text{GradL}\left(x,y,\lambda_1,\lambda_2\right) := \nabla_{x,y,\lambda_1,\lambda_2} L\left(x,y,\lambda_1,\lambda_2\right) \rightarrow \begin{bmatrix} \lambda_2 + \dfrac{1}{\left(x^2 + 5\right)\cdot\left(y^2 + 5\right)} - \dfrac{\lambda_1\cdot x}{2} - \dfrac{2\cdot x\cdot(x - y)}{\left(x^2 + 5\right)^2\cdot\left(y^2 + 5\right)} \\[3mm] \lambda_1 - \lambda_2 + \dfrac{y}{10000} - \dfrac{1}{\left(x^2 + 5\right)\cdot\left(y^2 + 5\right)} - \dfrac{2\cdot y\cdot(x - y)}{\left(x^2 + 5\right)\cdot\left(y^2 + 5\right)^2} \\[3mm] y - \dfrac{x^2}{4} \\[3mm] x - y + 1 \end{bmatrix}$$

$$x := 0 \qquad y := 0 \qquad \lambda_1 := 0 \qquad \lambda_2 := 0 \qquad \Leftarrow \text{ Initial guesses, required by solver}$$

Solve gradient equations to find location of minimum

$$\text{Given} \qquad \text{GradL}\left(x,y,\lambda_1,\lambda_2\right) = \begin{pmatrix} 0 \\ 0 \\ 0 \\ 0 \end{pmatrix} \qquad \begin{pmatrix} x_{star} \\ y_{star} \\ \lambda_{1star} \\ \lambda_{2star} \end{pmatrix} := \text{Find}\left(x,y,\lambda_1,\lambda_2\right) = \begin{pmatrix} -0.828 \\ 0.172 \\ 5.505 \times 10^{-3} \\ -0.027 \end{pmatrix}$$

Negative Lagrange multiplier shows that constraint g_2 is not active

Fig. 6.15 Finding the constrained minimum with two inequality constraints

References

1. Kuhn HW, Tucker AW (1951) Nonlinear programming, proceedings of the 2nd Berkeley symposium, p 481–492
2. Karush W (1939) Minima of functions of several variables with inequalities as side constraints, M. Sc. Dissertation, Dept. of Mathematics, Univ. of Chicago

Chapter 7
Discrete Variables

Data by itself is useless. Data is only useful if you apply it

-Todd Park

So far, all the problems have involved continuous variables. For example, the value of a resistor in a power circuit could be anything we choose. It's not a problem if you want a resistor with a nonstandard value like 123.45 Ω, though you might find that resistors with common values, like 121 Ω, are cheaper. However, assuming that you're willing to pay the cost, you can have any resistance you want; no physical law prevents it. However, not all design variables are like that.

Imagine that you are designing a cruise ship and you need to specify how many cabins it should have for paying customers. That number has to be an integer. You can't have, for instance, 512.5 cabins. The number of cabins on the ship is an example of a discrete variable.

Discrete problems are often very difficult to solve, and the methods developed for continuous variables, like steepest descent, are not generally suited to discrete problems. Sometimes, people attempt to treat discrete variables as if they are continuous and then just round the resulting values of the design variables, x^*, to the nearest integer. This rough approach can sometimes work well enough; however, it rather misses the point.

7.1   The Traveling Salesman Problem

Perhaps the most famous discrete problem is the traveling salesman problem [1]. It is a classic problem because it is easy to state, is devilishly hard to solve, and is representative of a whole class of more general discrete optimization problems. That is, if you can solve the traveling salesman problem, even approximately, you can apply that same method to many other useful problems.

The traveling salesman problem is very simple: imagine there are n cities, and a traveling salesman is required to visit them all on a single trip (in pre-Internet days,

© Springer International Publishing AG, part of Springer Nature 2018
M. French, *Fundamentals of Optimization*,
https://doi.org/10.1007/978-3-319-76192-3_7

Fig. 7.1 Traveling
salesman problem, $n = 5$

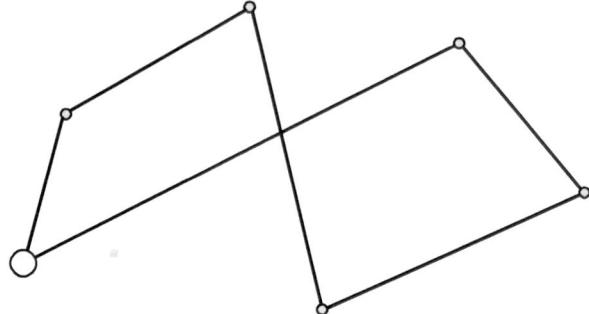

sales people apparently did this). Find the shortest path that visits each city once and returns to the origin. Figure 7.1 shows an example path for $n = 5$, where n designates the five cities that are not the starting point.

This is an extremely challenging problem because of how many possible paths must be evaluated in order to be sure you have found the minimum length path. Say there are four cities to be visited. The first step has four possible paths, the second step has three possible paths and so on. Thus, the number of possible paths is $N = 4 \times 3 \times 2 \times 1 = 4!$. In general, the number of possible paths is $n!$, where n is the number of cities to be visited. Factorials get big very quickly, so finding the minimum length path when $n = 10$ requires evaluating 3,628,800 paths – a big, but still manageable number. However, increasing problem size to $n = 20$ requires evaluating 2.433×10^{18} paths. That's 2.433 million trillion paths. Clearly, just calculating all the possibilities and selecting the shortest one is only possible for the smallest problems, and some more sophisticated approach is necessary. Because the variables must take discrete values, conventional search methods are difficult to apply. This basic intractability is the core of the problem.

If we are content with an approximate solution rather than an exact one, several possibilities present themselves. An approximate solution can be one that finds a point in design space at which the objective function is low, but not necessarily a global or even a local minimum. Generally, an approximate minimum is better than no answer at all, so some of these approximate solution methods have been widely used. A simple one is called the nearest neighbor (NN) algorithm.

7.2   Nearest Neighbor Algorithm

The nearest neighbor (NN) algorithm is based on the idea that it is not difficult to calculate the distance from a given point to all the other ones [2, 3]. NN just moves successively to the closest point. Starting at the origin, x_0, the first step is to the nearest point, which is then designated x_1. The second step is to the unvisited point closest to x_1 and this new point is designated as x_2. This process is repeated until no unvisited points remain. The last step is back to the origin. To illustrate, imagine a

Fig. 7.2 Illustration of the nearest neighbor algorithm, $n = 5$

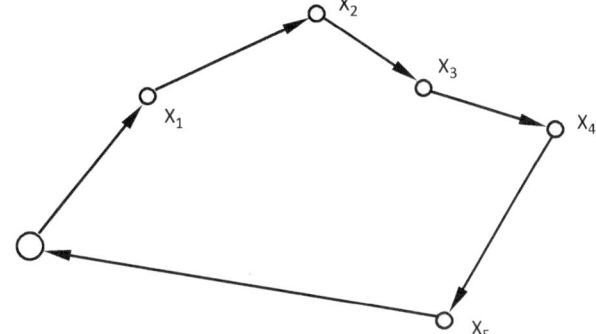

collection of points arranged so that the successive steps are obvious, as shown in Fig. 7.2. It is worth noting that the NN algorithm is an instance of a more general class called greedy algorithms that make the locally optimal choice at each step ignoring future consequences [4].

It's no surprise that the last step back to the origin is often a long one. That said, the method is easy to implement, has low computational cost, and can give acceptable answers. It is important to note that there is no guarantee that the solution will be close to the minimum. Occasionally, NN will produce a very poor path, one much longer than the minimum distance. This is the price that must be paid in order to find an approximate minimum for a problem where the exact solution may be impossible to compute.

A simple MATLAB program implementing the nearest neighbor algorithm shows the advantage of accepting an approximate minimum rather than insisting on a global minimum. In the first example here, the origin is at point (0,0), and there are eight randomly placed points to be visited. For convenience, the dimensions are normalized so that $0 < x < 1$ and $0 < y < 1$. Figure 7.3 shows a sample arrangement of points along with the global minimum and the one found using NN.

The nearest neighbor (NN) algorithm found an answer within about 7% of the global minimum. Since n is small, it was possible to calculate every possible path and to simply find the global minimum that way. In looking through a handful of other examples for $n = 8$, NN found the global minimum once and was never farther than 25% from the global minimum. The algorithm is very simple and works reasonably well for small n; however a more interesting story comes from computational time.

Even allowing for the fact that the author might not be the most efficient programmer, the time required for NN to run is remarkably small compared to running every possible permutation. Simply using the "tictoc" timing function in MATLAB for the $n = 8$ case shown here, NN ran in 0.163 s while computing every possible path took 4.174 s – a time savings by NN of about 96%. The cost of evaluating every possible route gets big very fast as n increases, as n! dictates. For $n = 9$, computing every possible path took 576 s for the random case examined.

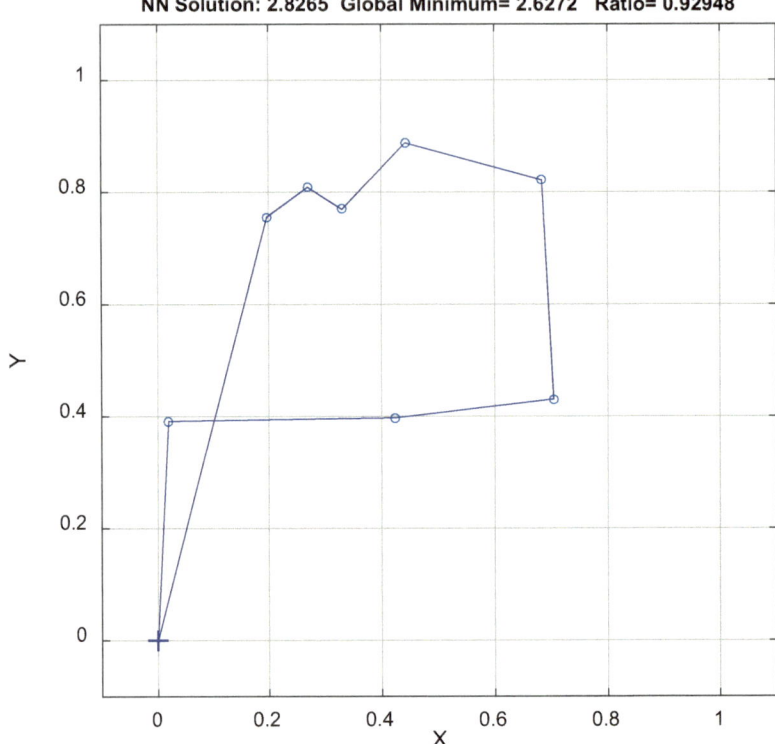

Fig. 7.3 Traveling salesman problem, $n = 8$

Increasing problem size to $n = 10$ gave an unacceptably long computation time. Note that these calculations were made on a small laptop computer in 2016 – please make allowances for hardware improvements as you read this and try not to snicker at what will soon enough seem primitive hardware.

If my little computer can only reasonably calculate every possible path for $n = 9$ or $n = 10$, problems much bigger than that are completely unrealistic. Let's see what NN does when $n = 100$ and the number of possible paths is $100! = 9.333 \times 10^{157}$. This is a mind-bendingly huge number. If my computer requires a microsecond (0.000001 s) to evaluate each possible path, it would take about 9.333×10^{151} s to evaluate every one of them. That's 2.958×10^{144} years, while the age of the observable universe is currently estimated to be about 1.378×10^{10} years.

While calculating the global minimum for $n = 100$ is absolutely hopeless, the NN algorithm can find an approximate minimum without trouble. Figure 7.4 shows the path calculated for 100 cities to be visited. The obvious question is how to be sure that the calculated distance of 9.7223 is close to the global minimum. The short answer is that there is no way to be sure, since the number of possible paths is so astronomically large as to be incomputable.

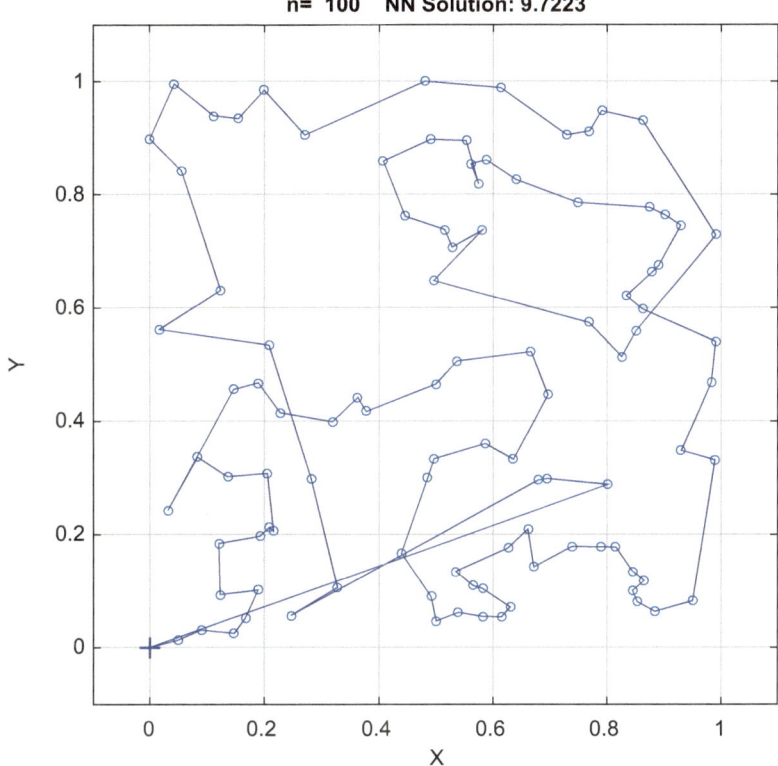

Fig. 7.4   NN approximate solution to the traveling salesman problem, $n = 100$

The traveling salesman problem and variations on it have been the object of extensive study, and there is a whole body of literature on the subject. It is no surprise, then, that there are other algorithms for generating approximate solutions. The NN algorithm presented here is probably the simplest among them, and that makes it a good starting point for this problem. Other, more sophisticated methods, like simulated annealing [5] and ant colony optimization [6], can give generally better results but with a more complex algorithm. Before leaving the problem, let's go to the extreme and try $n = 500$. For the record, MATLAB returns infinity when asked to calculate 500!, though Wolfram Alpha returns 1.2201×10^{1134}. This number is hundreds of orders of magnitude larger than any useful value and is far too large to even have a physical meaning. For example, the Eddington number, N_{edd}, the number of protons in the visible universe, is sometimes estimated to be 10^{89}. In spite of the number of possible paths, NN calculates an approximate path, shown in Fig. 7.5, very quickly.

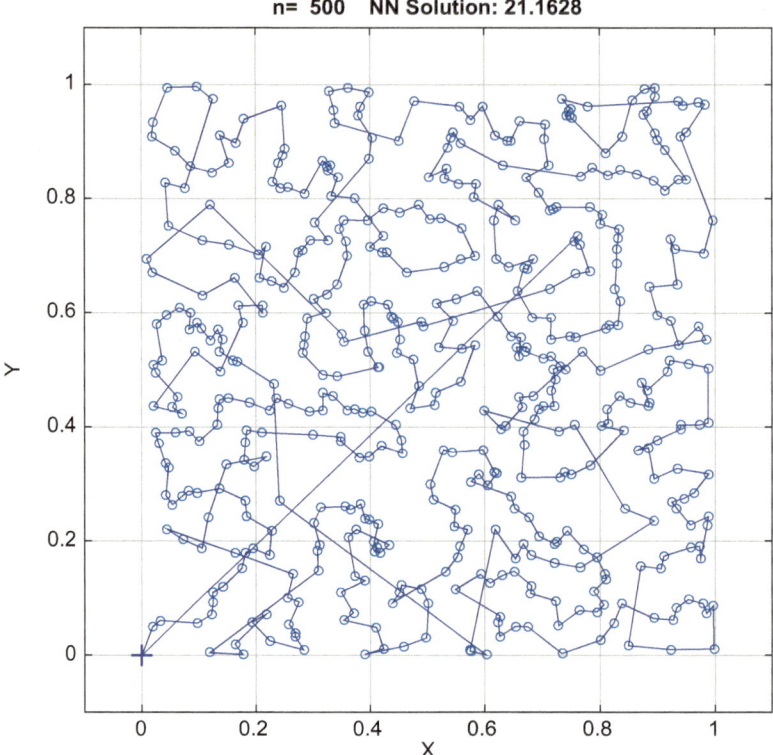

Fig. 7.5 NN approximate solution to the traveling salesman problem, $n = 500$

7.3 Applications of the Traveling Salesman Problem

The traveling salesman problem is certainly idealized in that it describes a problem reminiscent of an earlier age. The advent of the communication age means that salesmen are much less likely to "work a territory" by visiting a succession of customers over the course of long trips. It is also (happily) less likely that these sales people are men. Yet, this problem is closely related mathematically to other problems of great practical interest. Two of these are vehicle routing and micropro-cessor chip design,

7.3.1 The Vehicle Routing Problem

The vehicle routing problem (VRP) is a generalization of the traveling salesman problem [7]. In its simplest form, VRP answers the question "What is the optimal collection of routes that must be traveled to deliver goods to a group of customers?"

The definition of optimal depends on the context of the problem. It could mean the shortest cumulative distance, the shortest cumulative time, the lowest net cost, or something else.

As we move from shopping in stores to shopping online, the problem of delivering packages becomes central to national economies. As this is being written, delivery trucks belonging to delivery companies are ubiquitous as people shopping online want their orders delivered quickly – often in a day or two. Amazon is now experimenting with delivering parcels using autonomous vehicles. As you read this, Amazon's experiment may seem either prescient or naïve, depending on how well it works.

Consider the case of a package delivery service in an urban area, like Chicago (think FedEx or UPS). There may be a fleet of trucks delivering thousands of packages per day to hundreds or thousands of locations. The number of possible routes is certainly huge, but there are other complicating factors. The packages must probably be divided up among the trucks in such a way that no recipient is visited by more than one truck. Some routes may be undesirable because of construction or heavy traffic. Even turns might matter. One delivery service in the USA determined that left turns across traffic took so much extra time, on average, that they were banned (if your country drives on the left, the equivalent problem would, of course, be right turns).

This touches on what may be the most challenging part of the problem – the delivery trucks must follow roads. They cannot take more direct routes as could helicopters or drones. In fact, it's probably worse than that, since some roads are one way and not all roads are accessible from one another; if you are driving on a highway, you can only leave it using exit ramps. Figure 7.6 shows a part of the Chicago road system. A big package delivery company could easily be required to deliver many thousands of packages per day in an area this large.

7.3.2 Computer Circuit Design

Nearer to the other end of the distance scale, let's look at computer wiring. The earliest computers were so slow that the time it took signals to move from place to place within the machine was small compared to the time it took the machine to actually do anything. Figure 7.7 shows the ENIAC, an early digital computer that ran using vacuum tubes rather than transistors, which were not yet available [8].

As computers got faster, the speed at which electromagnetic signals propagate through wires began to actually matter, and designers had to be more careful in how connections were made within their machines. A well-known series of supercomputers were manufactured by Cray computers. They were surprisingly small, given their power. And, while they were tidy, even elegant on the outside, the interiors were packed with wire. Figure 7.8 shows wiring on the interior of a Cray-2.

The early generations of supercomputers, like those from Cray, usually had a small number of very fast (for the time) processors. Thus the challenge for the

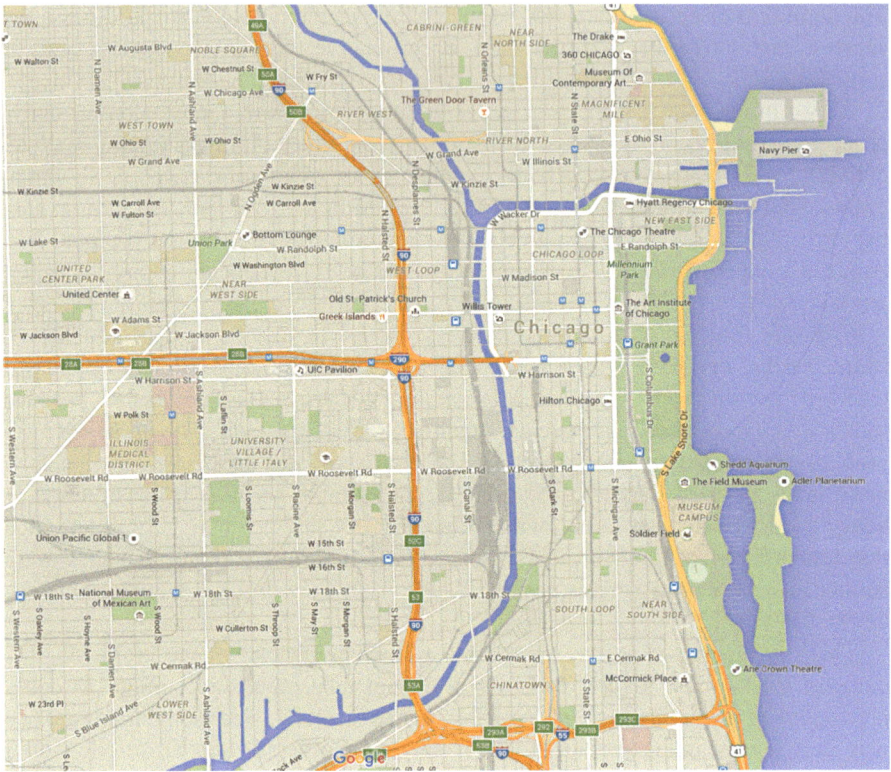

Fig. 7.6 A map of a portion of the Chicago road system. (Google Maps)

designers was to find ways to make those processors faster. One way to do that was to pay attention to the wiring runs inside the machine. This is a great example of discrete variables because the wires had to run between fixed points and each had to take one of a finite, though large, number of possible routes. To cast the problem in terms of optimization, the objective was to minimize the time to make a calculation (or perhaps a representative group of calculations) – calculation time is the objective function. The design variables are the paths the wires might take in order to form the necessary circuits.

The author had the opportunity in the late 1980s to visit a Cray facility in Minnesota. One feature of the machines on site was that the disk drives were on the floor above the central processing units (CPUs). The tour guide explained that placing the disks one floor above the CPUs meant that, on average they were closer than if they had placed on the same floor. That slight reduction in distance mattered. Again, that decision involved an integer design variable. The disk drives had to either be on the same floor as the CPU or on the next floor up. They couldn't go halfway between.

Perhaps, the designers of the building could have given the room a high ceiling so that the disks could be mounted on a frame right above the CPU. However, that option was apparently rejected. One might also assume the disks could have been

Fig. 7.7 The ENIAC computer showing external wiring. (Wikimedia Commons, image is in the public domain)

mounted on the floor below the CPUs, but this would seem to be no better than mounting them on the floor above.

As computing power increased, along with clock speed, the time required to push signals through the circuitry became a matter of the foremost importance. In response, more of the components were placed on the same silicon chip as the processor. Figure 7.9 shows the die for a Motorola 68,040. While this is, even as I write this, an old design, it clearly shows the problem faced by chip designers.

The chip is essentially two dimensional, so all the components and their interconnections have to be laid out on a surface (though it should be noted that multilayer chips are now in use). The similarity to the TSP and the VRP may be intuitive at this point. The designers have to lay out the components and their connections to minimize travel time of signals while still satisfying other design constraints. As with the other problems, the design variables are discrete – the various components cannot overlap one another, and the interconnections must be made between given points, not some arbitrary point in between.

Fig. 7.8 Wiring bundles inside a Cray-2 supercomputer. (Wikimedia Commons, image is in the public domain)

7.4 Discrete Variable Problems

Many of the methods used for continuous problems depend on slopes and, sometimes, curvatures. However, the concept of slope doesn't exist for discrete variables. As an example, consider the objective function

$$f(x,y) = \frac{e^{-[\text{round}(x)/6]^2}}{[\text{round}(y) + \tan^{-1}(\text{round}(x) + 1)]^2 + 4} \tag{7.1}$$

where "round" indicates rounding the variable value to the nearest integer. Thus $f(x, y)$ is a function of integer values of x and y, but $f(x, y)$ is not restricted to integers. Figure 7.10 shows the function as a 3D bar graph. It is clear from this figure that the slope can only be defined approximately by fitting a curve or a surface over several bars.

While there are many discrete methods in the literature, a good start is the hill climbing method [9, 10]. It is simple to implement and can reliably get close to local minima. The problem here is defined as a maximum rather than a minimum simply because the hill climbing method is, as the name suggests, intended to find the

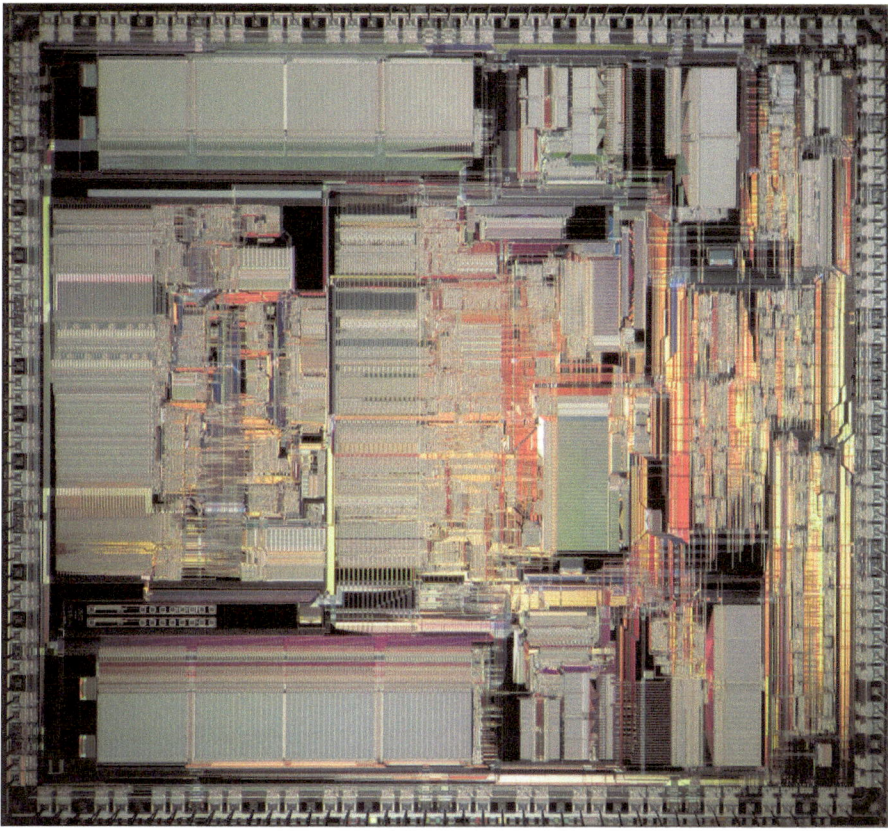

Fig. 7.9 The die for a Motorola 68040 microprocessor. (Wikimedia Commons, image is in the public domain)

maxima. Finding a minimum of a function is the same as finding the maximum of the negative of the function, that is,

$$\max(f) = \min(-f) \tag{7.2}$$

There are a number of variations on the hill climbing method in the literature, and the method is simple enough to lend itself readily to modification. The version described here works well, but it is certainly not the only one.

The big idea is that, from a given starting point, the height of adjacent bars or steps of the objective function are compared with the current value of f. If the new f is greater than the current one, then x or y is indexed to form the new estimate of the maximum.

In the example here, x is varied first; if adding 1 to or subtracting 1 from index of x increases the objective function, then the estimated location of the maximum is updated. Then, the y index is varied by ± 1 and the estimate is updated if necessary.

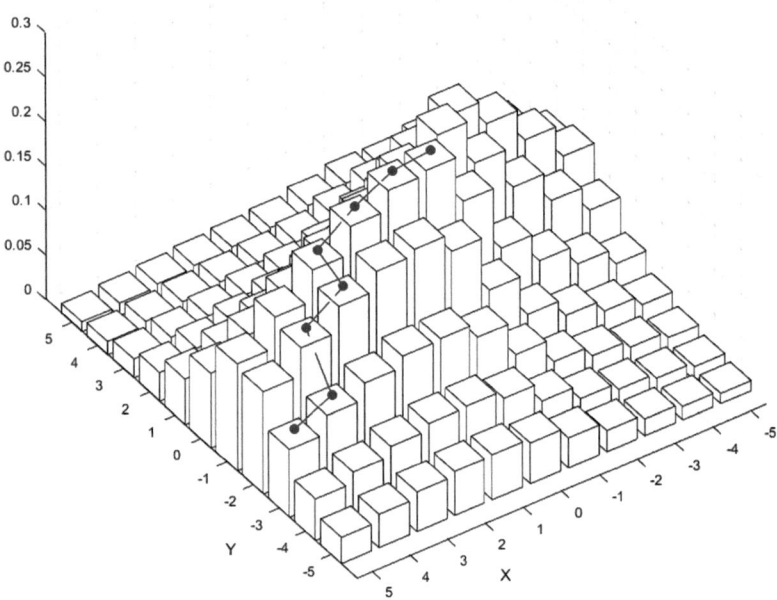

Fig. 7.10 3D bar graph of the discrete test function

The process repeats until neither unit changes in x nor in y increase the objective function (Fig. 7.11).

One of the limits of this version of the hill climbing method is that it can only move in the X or Y direction. To address this, it would be possible to evaluate variations in x and y simultaneously. With this modification, hill climbing starts to resemble a discrete version of marching grid.

Finally, Fig. 7.12 shows a quasi-flowchart for the hill climbing method in two variables, x and y. The method is easily extended to an arbitrary number of variables.

7.5 Examples of Discrete Variable Design Problems

Discrete design variables are common in the design process, and they often present themselves in ways that are not directly related to the traveling salesman problem. For example, structural materials and fasteners are often available in a limited number of common sizes, and wire is available in a fixed number of common gauges. Even more fundamental are preliminary design decisions like how many processors should operate a computer or how many engines an airliner should have. What follows are two of many possible examples of discrete design variables.

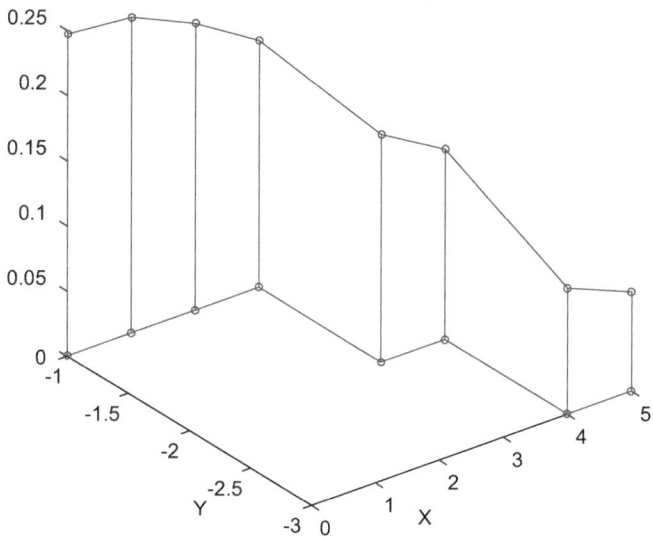

Fig. 7.11 Example solution path for hill climbing method, starting point $(5, -3)$

Define starting point, x_0, y_0
Define discrete changes in x and y, Δx and Δy
$n=0$

If $f(x_n-\Delta x) > f(x_n)$, then $x_{n+1} = x_n-\Delta x$
Else if $f(x_n+\Delta x) > f(x_n-\Delta x)$ then $x_{n+1} = x_n-\Delta x$
Else $x_{n+1}=x_n$
If $f(y_n-\Delta y) > f(y_n)$, then $y_{n+1} = y_n-\Delta y$
Else if $f(y_n+\Delta y) > f(y_n-\Delta y)$ then $y_{n+1} = y_n-\Delta y$
Else $y_{n+1}=y_n$
If $x_{n+1}=x_n$ and $y_{n+1}=y_n$ ⟶ Stop
$n=n+1$

Fig. 7.12 Quasi-flowchart for the hill climbing method

7.5.1 Marine Diesel Engine

As this is written, I am aboard a ship powered by four very large diesel engines. They are made by MAN and are model 12 V48/60. This designation means that they are V12 (12 cylinders in a V arrangement connected to a single crankshaft), with a bore of 48 cm and a stroke of 60 cm. This gives a displacement of about 1300 L. Note that the engine on a small car in the USA is typically about 2 L and the large engine in an American "muscle car" might be in the neighborhood of 6 L. The physical size of the

Fig. 7.13 A MAN 12v48/60 marine diesel engine. (Image from MAN SE)

engine is clear in Fig. 7.13. Note the walkway and handrails around the upper part of the engine. This ship has four of these engines to power the propellers, to provide electricity for the ship, and to distill freshwater from seawater.

These engines consume a large amount of fuel, enough that the ship's bunkers hold about 2000 metric tons. Clearly, the engines must be efficient and reliable while not being prohibitively expensive to purchase or to operate. One of the most fundamental decisions to be made by the engine designers is the number of cylinders. Clearly, the number of cylinders has to be an integer. In practice, it is usually an even integer.

More cylinders give more displacement and, thus, more power but add length and complication. Conversely, a smaller number of larger cylinders reduces the complexity but requires making larger and, likely, more expensive parts.

It could be cheaper to make a family of engines that all use the same bore and stroke so that major components are common to all engines in the family. Indeed, this appears to be what the designers at MAN have done as the engine is available in inline 9, V12, V14, and V18 configurations. Using these basic configurations, individual customer needs can be accommodated using a selection of common parts. The manufacturer lists power at 500 rpm as 1.05 MW per cylinder.

It is interesting to note that the heat carried away from the engine by the coolant, usually just rejected to the environment and wasted, can be used to distill freshwater from seawater. Fresh water is always a valuable commodity aboard a ship, and being able to make it for free simply by running ocean water in the outer coolant loop through a distiller reduces both cost of operating the ship – freshwater does not always have to be purchased while in port – and reduces the need to accommodate freshwater storage tanks large enough to support an entire voyage.

Of course, the cylinder configuration of the engines is just the beginning of the discrete design problems. One must also decide on the number of engines. The ship on which I traveled had four engines, each driving a generator. They provided electrical power for both the activities on the ship and the propulsion system, which was built around two large electric motors and three smaller ones. The ship generally used between two and four engines, depending on desired speed and the electrical load.

The designers could have selected any number of engines, but apparently decided that four would be both reliable and efficient. On the return part of the voyage, the ship was running on two or three engines while the fourth was undergoing maintenance. A ship in port is not making money for the owner and this engine configuration allowed major maintenance to take place while the ship was underway.

Looking farther into this design problem, there are still more discrete optimization problems. Since the screws (also called propellers, but everything seems to get a new name when it is put on a ship) are driven by electric motors, there is no automatic relationship between the number of engines and the number of screws. On this ship, there were four engines and two screws, though there were also three side thrusters. Next, there is the decision of how many blades should be on each screw, and so on.

7.5.2 Wind Turbines

Windmills have been used for mechanical energy for centuries, and wind turbines have been used to make electrical power for decades. As wind turbines developed, designers sought to make them as efficient as possible given material limitations. Since wind turbines just turn kinetic energy of moving air into electrical energy, efficiency can be defined as the fraction of kinetic energy converted. Completely stopping the moving air would, thus, give an efficiency of 1.00 or 100%.

The obvious problem is that the now motionless air would block any more air from moving past the turbine blades. There is an upper limit to the energy that can be extracted from moving air by wind turbine. This maximum, called the Betz limit [11], is 59.3% for turbines driven by aerodynamic lift (the limit is considerably lower

Fig. 7.14 NASA/DOE Mod 5B wind turbine. (Wikimedia Commons, image is in the public domain)

for drag turbines). That is, no lift turbine can extract more than 59.3% of the kinetic energy from moving air while in continuous operation.

So a possible design goal is to make a wind turbine capable of operating as close to the Betz limit as possible. One of the many design choices is how many blades the turbine should have. Again, this is clearly a discrete design variable since the number of blades has to be an integer. The mechanical and aerodynamic considerations are complicated, so the right answer was not obvious to the designers of early wind turbines.

Starting in the 1970s, NASA and the Department of Energy (DOE) sponsored a series of experimental wind turbines to evaluate a range of possible design features. All had two blades, though they were of very different sizes. The last design, designated Mod 5B, had a blade diameter of 100 m and was capable of generating 3.2 MW. Figure 7.14 shows this turbine installed in Hawaii.

Fig. 7.15 Offshore wind turbines. (Wikimedia Commons, image is in the public domain)

While these turbines developed technologies and introduced design features, subsequent large commercial wind turbines almost all have three blades. Figure 7.15 shows three large turbines in an offshore installation.

References

1. Cook WJ (2014) In pursuit of the traveling salesman: mathematics at the limits of computation. Princeton University Press, Princeton
2. Guton G, Yeo A, Zverovich A (2002) Traveling salesman should not be greedy: domination analysis of greedy-type heuristics for the TSP. Discret Appl Math 117:81–86

3. Reinelt G (1994) The traveling salesman problem: computational solutions for TSP applications, springer lecture notes in computer science 840. Springer, Berlin

4. Hazewinkel M (2001) Greedy algorithm, encyclopedia of mathematics. Springer/Kluwer, Netherlands

5. Kirkpatrick S, Gelatt CD Jr, Vecchi MP (1983) Optimization by simulated annealing. Science 220:671–680

6. Dorigo M, Maniezzo V, Colorni A (1996) Ant system: optimization by a Colony of cooperating agents. IEEE Trans Syst Man Cybern B 26(1):29–41

7. Laporte G (2009) Fifty years of vehicle routing. Transp Sci 43(4):408–416

8. Brainerd JG, Sharpless TK (1999) The ENIAC. Proc IEEE 87(6):1031–1041

9. James PN, Souter P, Dixon DC (1974) A comparison of parameter estimation algorithms for discrete systems. Chem Eng Sci 29:539–547

10. Tovey CA (1985) Hill climbing with multiple local minima. SIAM J Algebraic Discrete Methods 6(3)

11. Betz A (1966) Introduction to the theory of flow machines. Pergamon Press, Oxford

Chapter 8
Aerospace Applications

There is an art or, rather a knack, to flying. The knack lies in learning how to throw yourself at the ground and miss.

-Douglas Adams

There are many applications of optimization in the aerospace world. The natural design tradeoffs inherent in designing an airplane require the designer to think about what the objective function needs to be and what constraints need to be applied. In many cases, the resulting airplane is intended to serve in multiple roles, but sometimes the objective function is so important that it dominates the design. This can be seen in some of the more extreme designs.

One extreme is an airplane designed to be very fast. Flying at slow speeds, as required for takeoff and landing, requires a large wing while flying at high speeds doesn't. In fact, the large wing required for takeoff and landing generates unwanted drag at high speed. An example of an airplane optimized for high speed is the F-104 shown in Fig. 8.1.

The plane was optimized to fly very fast and could reach speeds in excess of Mach 2 [1]. However, its takeoff and landing speeds were high, and it developed a reputation as a demanding airplane to fly. It had relatively little interior volume and was often flown with external fuel tanks (like the ones shown here) that could be jettisoned when empty to reduce drag. In practice, the plane proved to be so specialized for high speed flight that it was poorly suited to the wide range of other missions needed by most air forces. While it was purchased by several different countries in NATO, accident rates were very high, and it was not considered an overall success.

On the other end of the spectrum, let's consider aircraft designed to be extremely efficient at low speeds. Sailplanes usually have no engines and remain aloft by circling in rising air currents called thermals [2]. In order to work, they need very low minimum sink rates and very low drag. This requires minimizing power required to stay aloft at low speeds. A representative sailplane is shown in Fig. 8.2.

The wings have a very high aspect ratio, and the entire airframe is designed to have low drag at low speed. The result is a plane that handles well at low speed and

M. French, *Fundamentals of Optimization*,
https://doi.org/10.1007/978-3-319-76192-3_8

Fig. 8.1 Lockheed F-104. (Wikimedia Commons, image is in the public domain)

Fig. 8.2 A sailplane designed for high efficiency at low speed. (Wikimedia Commons, image is in the public domain)

has a minimum sink rate so low that it can gain altitude by circling in thermals (rising masses of warm air).

However, the plane is so optimized that it is poorly suited to any other task. It has no engine so it must be towed into the air by another plane or launched using a winch. Some sailplanes have small retractable engines or electric motors in order to launch themselves, but they are generally used only for takeoff and climb. The sailplane shown here has room only for the pilot, who is essentially laying down so the fuselage can have a small cross section. To reduce weight and drag, it doesn't even have conventional landing gear. Rather it has a single wheel near the center of gravity.

In the following few sections, let's explore in more detail how optimization has been used in aircraft design, starting with wing configuration.

8.1 Monoplane or Biplane

In the early days of aircraft design, one of the important design considerations was whether an airplane should have one set of wings or two. That is, should it be a monoplane or a biplane? The earliest airplanes were generally made of wood with the structures stiffened by diagonally oriented wires. The wings were very thin in cross section, and external wire stiffening was required to make them strong enough to bear the loads imposed by violent maneuvering [3].

Figure 8.3 shows a replica of a Sopwith Camel, a very successful fighter aircraft from WWI. The wings are very thin and could not withstand maneuvering loads without the external wire bracing shown in this picture.

Designers were well aware of the additional drag created by the additional wing and the wire bracing. However, this arrangement resulted in a very strong, light structure. The strength of a structure is a strong function of its depth. In this case, the depth of the structure is the distance between the wings.

By the end of WWI, monoplanes were just starting to enter wide use. Designers developed thicker airfoils that allowed the wing structure to be placed largely inside the wing. Perhaps the most structurally sophisticated aircraft of WWI was the Fokker D-VIII. The wing was much thicker than was then common, so a strong structure could be completely enclosed by the aerodynamic envelope.

Figure 8.4 shows a replica D-VIII with a complete absence of external bracing wires. The only external structure was the steel tubing that supported the wing above the fuselage. This arrangement reduced aerodynamic drag and helped to make it 10% faster (204 km/hr vs. 185 km/hr) than the Sopwith Camel.

Figure 8.5 shows a very interesting picture of a D-VIII with people seated and standing along the full length of the wing. This photograph was taken to demonstrate the strength of the aircraft's structure even though it was a monoplane without external wire bracing.

Fig. 8.3 A replica of Sopwith Camel. (Wikimedia Commons, image is in the public domain)

Fig. 8.4 A replica Fokker D-VIII. (Wikimedia Commons, image is in the public domain)

Fig. 8.5 Demonstrating the strength of the D-VIII wing. (Original source unknown)

Fig. 8.6 Elliptical wing
planform

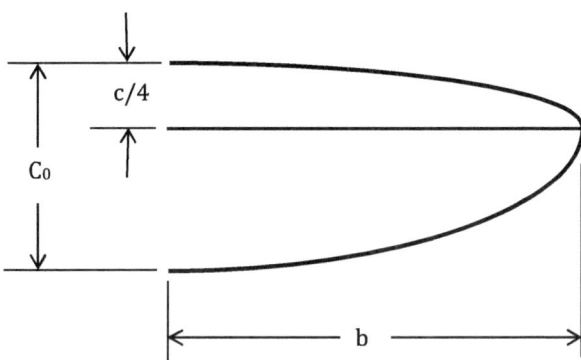

Biplanes are now almost completely obsolete. A few are still produced as aerobatic aircraft for which structural strength is extremely important and aerodynamic drag is of less concern.

8.2 Wing Planforms

The planform of a wing is just its shape when viewed from above or below. Wing planform plays an important role in the performance of the aircraft, and the literature on selecting planforms is vast. The story of how planforms evolved is a great example of optimization [4].

The amount of drag produced by a wing per unit lift is inversely proportional to a quantity called the planform efficiency factor, e. Basic low speed aerodynamic

Fig. 8.7 A Piper Cub with an untapered wing. (Wikimedia Commons, image is in the public domain)

analysis predicts an efficiency factor with a maximum value of 1.0 for an elliptical planform as shown in Fig. 8.6. It is one task of the designer to maximize the planform efficiency factor while meeting cost and weight constraints.

The least efficient of commonly used planforms is a straight, untapered wing as shown on the Piper Cub in Fig. 8.7. It has the advantage of having a constant cross section, so all the internal structure is simple. So, while it is less efficient, it is much easier to make and, thus, costs less. This planform also contributes to benign handling characteristics. An old joke in the aviation world is that the Cub is the safest plane there is – it can just barely kill you.

Tapering all or part of the wing can give a rough approximation to an elliptical planform and is more efficient than a rectangular planform. Figure 8.8 shows the planform of a Cessna 152.

The inner portion of the wing is constant chord and the outer section is tapered. The result is a planform that is a rough approximation to an ellipse but much easier to build.

An elliptical wing is difficult to manufacture since there are often no straight lines on it anywhere. However, some designers believed that the additional complexity was more than offset by the increased efficiency. Perhaps the most famous example of an airplane with an elliptical wing is the Supermarine Spitfire as shown in Fig. 8.9.

This is all fine, but the calculation that results in the elliptical wing with a straight quarter chord line is the result of many simplifying assumptions. It is not hard to notice that birds don't have elliptical wings and, with few exceptions, fish don't have

Fig. 8.8 Planform of a Cessna 152. (Wikimedia Commons, image is in the public domain)

Fig. 8.9 A Spitfire showing the elliptical wings. (Wikimedia Commons, image is in the public domain)

Fig. 8.10 An ocean sunfish. (Wikimedia Commons, image is in the public domain)

elliptical fins. Either nature has been iterating for the last few billion years on the wrong solution or there is more to the problem than can be explained using simple aerodynamic analysis.

There is as least one fish with approximately elliptical fins, the ocean sunfish. Figure 8.10 shows an ocean sunfish with the dorsal and ventral fins clearly visible.

A much more familiar shape is that of the tuna as shown in Fig. 8.11. It is a very fast fish, but neither the fins nor the tail is elliptical. More sophisticated aerodynamic analysis has shown that a wing with slightly swept tips (sometimes called a lunate or scimitar planform) is slightly more efficient than a straight elliptical wing. This planform is now used in applications for which efficiency is particularly important.

Figure 8.12 shows a self-launching sailplane with a trailing edge that is nearly straight.

In a more clear application of swept planforms, Fig. 8.13 shows advanced, swept blades on the engines of a C-130 transport aircraft. We should note that another reason for the swept tips is to reduce transonic drag. While the plane itself flies at relatively low speeds, the tips of large spinning propellers can approach or even exceed the speed of sound.

Figure 8.14 shows the propellers on an older C-130 with broad, almost rectangular blades. This configuration was typical of propellers on turboprop aircraft since at least the late 1950s.

Fig. 8.11 Bluefin tuna. (Wikimedia Commons, image is in the public domain)

Fig. 8.12 A motorized sailplane with an efficient planform. (Wikimedia Commons, image is in the public domain)

Fig. 8.13 Scimitar-shaped propeller blades on a C-130. (Wikimedia Commons, image is in the public domain)

8.3 Vehicle Performance

Optimization is central to the development of expressions describing aircraft performance. For example, the maximum rate of climb occurs at the speed for which power required to maintain level flight is a minimum [4]. The development of the expression for power required is not difficult and is worth showing here.

Power required to maintain level flight is the product of drag and velocity

$$P_{\text{req}} = \text{Drag} \times \text{Velocity} = C_{\text{D}}\frac{1}{2}\rho v^2 s \times v = C_{\text{D}}\frac{1}{2}\rho v^3 s \tag{8.1}$$

where C_{D} is a non-dimensional drag coefficient, ρ is air density, v is velocity, and s is wing area. The expression relating lift and drag is

$$C_{\text{D}} = C_{\text{D0}} + \frac{1}{\pi e \text{AR}} C_{\text{L}}^2 \tag{8.2}$$

where C_{D0} is the drag coefficient at zero lift, e is the planform efficiency factor, and AR is the aspect ratio. The next step is to relate velocity to the lift coefficient. The expression for lift is

$$L = C_{\text{L}}\frac{1}{2}\rho v^2 s \tag{8.3}$$

Fig. 8.14 A C-130 with older, paddle blade propellers. (Wikimedia Commons, image is in the public domain)

where C_L is a non-dimensional lift coefficient that is a function of angle of attack. Since lift must equal weight, this expression can be rearranged to give

$$C_L = \frac{2w}{\rho v^2 s} \tag{8.4}$$

So, the expression for the drag coefficient becomes

$$C_D = C_{D0} + \frac{1}{\pi e \text{AR}} \left[\frac{2w}{\rho v^2 s} \right]^2 \tag{8.5}$$

And the expression for power becomes

$$P_{\text{req}} = \left(C_{D0} + \frac{1}{\pi e \text{AR}} \left[\frac{2w}{\rho v^2 s} \right] \right) \frac{1}{2} \rho v^3 s \tag{8.6}$$

Let's imagine an airplane with the following characteristics:

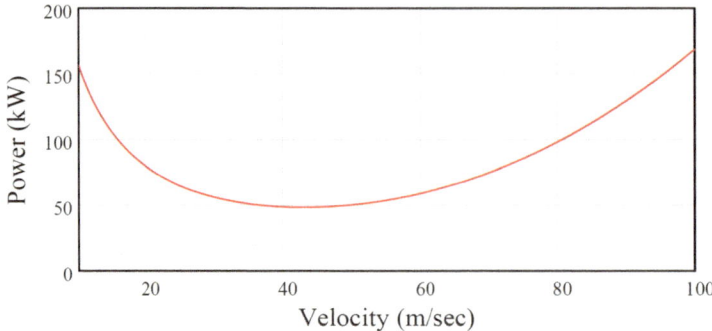

Fig. 8.15 Power required as a function of velocity

Fig. 8.16 Helios solar-electric aircraft in flight near the Hawaiian Islands (NASA)

$$C_{D0} = 0.017$$
$$e = 0.85$$
$$AR = 6$$
$$\rho = 1.2\,\text{kg/m}^3$$
$$s = 15\,\text{m}^2$$
$$w = 15000\text{N}$$

The plot of power required vs. velocity is shown in Fig. 8.15. The minimum power required is 48.5 kW and occurs at 42.94 m/sec.

8.4 Aircraft Design Optimization

A step forward in aircraft design has been the development of integrated software packages that optimize according to some objective function (often weight) and a collection of constraints [5]. The constraints may describe maximum stresses, aeroelastic stability, or other necessary properties.

The result has been the ability to optimize the design of aircraft with unusual configurations [6] or in difficult flight regimes [7, 8]. A notable example of this capability is the Helios, an unmanned electric aircraft shown in Fig. 8.16. It is powered by solar cells driving electric motors and demonstrated the ability to fly for extended times and very high altitudes. Its design was the result of extensive analysis and optimization, and the result was a plane that operated near the limits of what was then possible.

References

1. Dobrzynski J (2015) Lockheed F-104 Starfighter. Yellow Series, Hampshire
2. Thomas F, Milgram J (eds) (1999) Fundamentals of sailplane design. College Park Press, College Park
3. Grey CG (1990) Jane's fighting aircraft of world war I. Studio Publishing, London
4. Anderson J (2015) Introduction to flight. McGraw-Hill, New York
5. Veley DE, Canfield RA, Khot NS (1987) Structural optimization using ASTROS and STARS software, proceedings of computer applications in structural engineering conference, ASCE, pp 315–325
6. Rasmussen CC, Canfield RA, Blair M (2009) Optimization process for configuration of a flexible joined wing. Struct Multidiscip Optim 37(3):265–277
7. Kolonay RM, Yang HTY (1998) Unsteady aeroelastic optimization in the transonic regime. J Aircr 35(1):60–68
8. Kolonay RM, Eastep FE (1998) Optimal scheduling of control surfaces on flexible wings to reduce induced drag. J Aircr 43(4):574–581

Chapter 9
Structural Optimization

Science can amuse and fascinate us all, but it is engineering that changes the world.

-Isaac Asimov

Ask of the steel, each strut and wire, what gave force and power.

-Joseph B. Strauss

One of the widest applications of optimization is that of designing minimum weight structures [1]. This is particularly true of aerospace structures. The whole point of flying a plane or launching a rocket is to deliver a payload. A rocket may deliver a useful satellite to orbit, or an airliner may deliver passengers to their destination. In any case, anything that is not payload, like fuel and structure, represents a cost to be minimized. This is particularly true for rockets, in which the payload may only represent a small portion – a few percent – of the takeoff weight of the vehicle.

In the extreme, consider the Saturn V that launched the Apollo mission to the moon. The takeoff weight of the vehicle was 2,970,000 kg (6,540,000 lb). It could put 140,000 kg (310,000 lb) into low Earth orbit, for a payload mass fraction of about 4.7%. However, 48,600 kg (107,100 lb) could be put into lunar transfer orbit, for a payload mass fraction of 1.6%. Finally, the command module that was eventually recovered had an initial mass of 5560 kg (12,250 lb). The command module that finally returned the astronauts to Earth represented a payload mass fraction of about 0.2%. Thus even a small reduction in the takeoff weight of the rocket could have greatly increased the payload available to the astronauts. Conversely, a small increase in weight might have almost eliminated the useful payload.

When a system operates this close to the limits of what is possible, optimization is particularly valuable and may greatly improve overall performance. As a historical note, consider the space shuttle. For the first two launches, the external tank that held the liquid oxygen and liquid hydrogen for the main engines was painted white to protect the insulation that covered the tank. Afterward, though, the paint was eliminated, saving several hundred kg and slightly increasing the payload. The bare insulation gave the tank its familiar rust orange color.

© Springer International Publishing AG, part of Springer Nature 2018
M. French, *Fundamentals of Optimization*,
https://doi.org/10.1007/978-3-319-76192-3_9

In most cases, aerospace structures are modeled using finite element analysis, so the discussion that follows is based on matrix notation. There is a wide choice of objective functions, but weight is the most common.

9.1 Truss Structures

The range of possible structures is vast. Fortunately, we can do a simple example that shows the basics of structural optimization. For almost all practical problems, the structure is described by a finite element model. In large-scale problems, it is not unusual to have thousands of degrees of freedom and hundreds of design variables. Fortunately, some basic ideas can be communicated with the simplest of models. Let's consider a two-element truss with two design variables as shown in Fig. 9.1.

This truss is simple enough to analyze using basic statics. Let's assume the material is aluminum with an allowable stress of 100 MPa and an elastic modulus of 69 GPa. The truss elements are assumed to be round, with a wall thickness of t. They both have a length of 1044.03 mm.

The objective function will be mass of the structure. Because mass is just the product of volume and density, minimizing mass is essentially equivalent to minimizing the volume of the structure.

The constraints in structural optimization problems are intended to keep the structure from failing. In this case, element 1 might fail because of tensile stress, while element 2 might fail from compressive stress or from buckling. As a reminder, the buckling load of a pin-jointed bar is

$$P_{cr} = \frac{\pi^2 EI}{L^2} \tag{9.1}$$

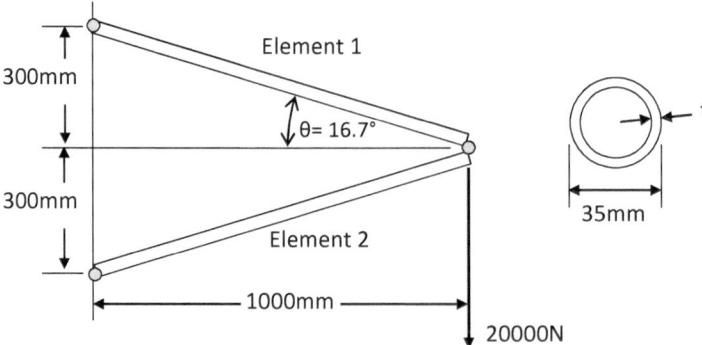

Fig. 9.1 Two-element truss

The last part of the setup is to identify the design variables. For now, let's assume we are using the same tube stock for both elements and use the wall thickness, t, as the design variable. Then, the formal problem statement is

Minimize   mass $= \rho^*$ volume

Subject to $g_1(t) = \sigma_1 - \sigma_{All} \le 0$ (tension)

$g_2(t) = -\sigma_{All} - \sigma_2 \le 0$ (compression)

$g_3(t) = -F_2 - \dfrac{\pi^2 EI}{L^2} \le 0$ (Buckling)

where σ_{All} is allowable stress, E is the elastic modulus of the material and I is the area moment of inertia.

The area moment of inertia of the tube is

$$I = \frac{\pi}{64}\left[D^4 - d^4\right] \tag{9.2}$$

For this problem, minimizing the structural volume means minimizing wall thickness, t. Since volume is a linear function of wall thickness, we are just trying to find the minimum wall thickness that satisfies all three constraints. Basic statics shows that $F_1 = 34{,}801$ N (tensile) and $F_2 = -34{,}801$ N (compressive), where F_1 and F_2 are axial forces. Stress is just force divided by cross-sectional area.

The cross-sectional area is

$$A = \frac{\pi}{4}\left[D^2 - d^2\right] \tag{9.3}$$

where D is the outer diameter and d is the inner diameter. Note that $d = D - 2t$. This simplifies to

$$A = \frac{\pi}{4}\left[D^2 - (D - 2t)^2\right] = \pi\left[Dt - t^2\right] \tag{9.4}$$

Since we're ignoring any additional mass at the joints, the total volume of the structure is 2LA or

$$V = 2LA = 2\,\pi L\left[Dt - t^2\right] \tag{9.5}$$

where D and t are both given in meters.

Let's start with the first constraint. The stress in element A is equal to the allowable stress when $t = 3.52$ mm. Since the magnitude of the forces in elements A and B is the same, this value of wall thickness satisfies both constraint 1 and constraint 2.

However, this wall thickness gives a buckling load of 27,271 N. Thus, the buckling load constraint is violated. The buckling constraint is only satisfied when $t = 5.20$ mm. This thickness will also satisfy the stress constraints. Thus the minimum weight structure is the one with $t = 5.20$ mm. The total volume of the structure is then 0.001017m^3. Figure 9.2 shows the constraint values for both the stress and buckling constraints.

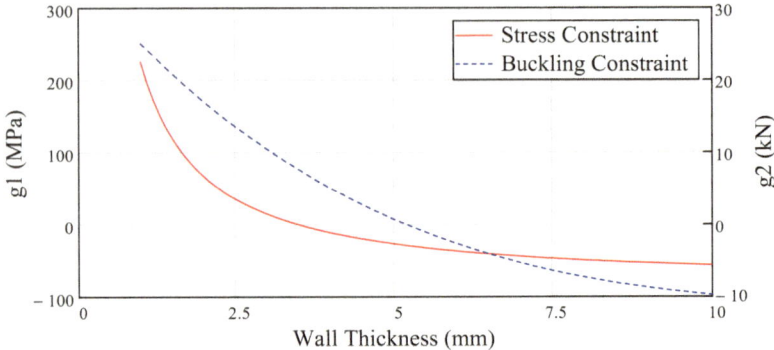

Fig. 9.2 Constraints for two-bar example problem

When the buckling constraint in element 2 is satisfied, element 1 is overdesigned. That is, element 1 is heavier than it needs to be in order to bear the tensile load. An obvious solution is to introduce an additional design variable by letting the two bars have different wall thicknesses. With this addition, element 1 can have a wall thickness of 3.52 mm, satisfying the tensile stress constraint. Element 2 would still have a wall thickness of 5.20 mm to satisfy the buckling constraint.

9.2 Updating Finite Element Models

Let's assume that a structure is undergoing a stiffness test. This is a test in which a known load is applied and the resulting strains and displacements are recorded. It is typical, for example, in the aerospace industry to run static structural tests on new aircraft designs as shown in Fig. 9.3.

Sensors placed at selected points on the structure generate data for comparison with finite element models. In cases where the model predictions do not match test data, the models are modified. Optimization-like processes are often used to determine how to modify (or update) the models so that predicted deflections, natural frequencies, and mode shapes more closely match measured ones.

The static equation of equilibrium for a finite element model is

$$\{f\} = [K]\{x\} \tag{9.6}$$

where $\{f\}$ is the vector of external forces, $[K]$ is the stiffness matrix, and $\{x\}$ is the vector of displacements.

It is possible to take derivatives of vectors and matrices just as with other mathematical entities. Let's say that some parameter of the model, a, is a design variable. This may be a wall thickness, a cross-sectional area, or some other physical parameter. It is easy to take the derivative of the equation of equilibrium with respect to a.

Fig. 9.3 Composite picture of A380 wing structural test. (Wikimedia Commons, image is in the public domain)

$$\left\{\frac{\partial f}{\partial a}\right\} = \left[\frac{\partial K}{\partial a}\right]\{x\} + [K]\left\{\frac{\partial x}{\partial a}\right\} \qquad (9.7)$$

The external forces are not usually functions of the design variables, so we can reasonably assume that $\{\partial f/\partial a\} = 0$ and

$$0 = \left[\frac{\partial K}{\partial a}\right]\{x\} + [K]\left\{\frac{\partial x}{\partial a}\right\} \qquad (9.8)$$

Solving for the change in displacement with respect to a gives

$$\left\{\frac{\partial x}{\partial a}\right\} = -[K]^{-1}\left[\frac{\partial K}{\partial a}\right]\{x\} \qquad (9.9)$$

This expression relates changes in the design variable to changes in the displacement vector. With this information, the design variables can be modified to reduce the difference between calculated and measured structural displacements.

This process can also be applied to dynamic testing. Let's assume that the first few resonant frequencies of a structure are determined experimentally and that they don't exactly match those from the finite element model. Tuning the model will require knowing how changing the design variable will change the eigenvalues of the model.

The eigenvalue equation is

$$[K]\{x\} = \lambda[M]\{x\} \tag{9.10}$$

Taking the derivative of this equation with respect to a, using the familiar product rule, gives

$$\left[\frac{\partial K}{\partial a}\right]\{x\} + [K]\left\{\frac{\partial x}{\partial a}\right\} = \frac{\partial \lambda}{\partial a}[M]\{x\} + \lambda\left[\frac{\partial M}{\partial a}\right]\{x\} + \lambda[M]\left\{\frac{\partial x}{\partial a}\right\} \tag{9.11}$$

This expression can be rearranged to give

$$[K]\left\{\frac{\partial x}{\partial a}\right\} - \lambda[M]\left\{\frac{\partial x}{\partial a}\right\} = \frac{\partial \lambda}{\partial a}[M]\{x\} + \lambda\left[\frac{\partial M}{\partial a}\right]\{x\} - \left[\frac{\partial K}{\partial a}\right]\{x\} \tag{9.12}$$

and pre-multiplied by $\{x\}^T$ to give

$$\left(\{x\}^T[K] - \lambda\{x\}^T[M]\right)\left\{\frac{\partial x}{\partial a}\right\} = \frac{\partial \lambda}{\partial a}\{x\}^T[M]\{x\} + \lambda\{x\}^T\left[\frac{\partial M}{\partial a}\right]\{x\}$$
$$- \{x\}^T\left[\frac{\partial K}{\partial a}\right]\{x\} \tag{9.13}$$

In almost all linear finite element models, $[K]$ and $[M]$ are symmetric, so

$$\{x\}^T[K] - \lambda\{x\}^T[M] = 0 \tag{9.14}$$

and

$$0 = \frac{\partial \lambda}{\partial a}\{x\}^T[M]\{x\} + \lambda\{x\}^T\left[\frac{\partial M}{\partial a}\right]\{x\} - \{x\}^T\left[\frac{\partial K}{\partial a}\right]\{x\} \tag{9.15}$$

The expression for the change in eigenvalues is

$$\frac{\partial \lambda}{\partial a} = \frac{-\lambda\{x\}^T\left[\frac{\partial M}{\partial a}\right]\{x\} + \{x\}^T\left[\frac{\partial K}{\partial a}\right]\{x\}}{\{x\}^T[M]\{x\}} \tag{9.16}$$

This expression predicts changes in eigenvalues with respect to changes in the design variables. This can be used to modify the finite element model so that the calculated resonant frequencies of a structure more closely match the calculated ones. A particular application of model updating is designing aeroelastically scaled wind tunnel models.

Fig. 9.4 A small test structure on an optical test table. (Image, the author and G.E. Maddux)

9.3   Aeroelastically Scaled Wind Tunnel Model Design

A special problem in designing aerospace vehicles is that of making the subscale models that can be tested in wind tunnels. If the goal is simply to determine the effects of air moving past the model, that model can be essentially rigid, with the structure being as heavy as needed.

However, some models are designed to be flexible so that the model duplicates, in scale, the dynamic interactions between the vehicle and the air through which it moves [2, 3]. Thus, the model must have specified stiffness response and specified dynamic response [4]. This is a type of inverse problem in which the response of a system is specified and the properties of the system must be determined. The problem is greatly complicated by the requirement that the resulting structure must fit within the scaled-down aerodynamic envelope of the vehicle.

Figure 9.4 shows a small experimental structure designed to have specified mass and stiffness properties for an aeroelastically scaled model. It is mounted in a heavy fixture on an optical test table. The laser and other optical components are for shooting double-exposure image-plane holograms that map deformation across the structure due to an enforced load or displacement.

When the need is great enough, entire aircraft can be the subject of aeroelastically scaled wind tunnel models. Figure 9.5 shows an F/A-18 E/F model mounted in the Transonic Dynamics Tunnel at the NASA Langley Research Center. The tunnel is unique in that it was designed to use either air or R-134a (normally used as a

Fig. 9.5 F/A-18 E/F flutter clearance model. (US Government, image is in the public domain)

refrigerant) and can test models in either fixed or free conditions. The model in this figure appears to have one or more rigid body degrees of freedom due to the cable supporting structure.

This is a sophisticated flutter clearance model developed to allow engineers to ensure the full-scale aircraft would be free of aeroelastic instabilities that could damage or destroy the plane in flight. Planes have been lost to aeroelastic instabilities since the early days of aviation [5], so this is no idle worry.

A simple way to approach the design problem is to decouple the stiffness and mass problem. In step one, the structure is designed so that it has the correct static response. There are different approaches to casting the stiffness design problem as an optimization problem. One is to cast the objective function as a squared error function between desired and calculated deformation due to a few loads. The design variables are the variable elements of the structure, such as beam widths or plate thicknesses.

In practice, it may not be necessary to include a large number of loads in the objective function. Displacement at a single point is usually the result of stiffness contributions from around the structure. Thus, it is increasingly unlikely that a structure having desired displacements due to a few widely separated point loads would give erroneous values when a few more loads are added.

The displacement is given by

$$\{x\} = [K]^{-1}\{f\} \tag{9.17}$$

For a concentrated force at a single degree of freedom, the static response is defined by a single column of $[K]^{-1}$. Thus, it is simple (if tedious) to validate portions of the global stiffness matrix experimentally.

In the second step, concentrated masses can be added to the structure to tune the resonant frequencies. These represent nonstructural mass terms that can be added to the global mass matrix. In practice, they are small weights added to the model at points corresponding to the grid points on the finite element model.

In this step, the structural masses are the design variables. The objective function might be a squared error function based on the desired and calculated resonant frequencies. The aeroelastically scaled model must have correct mode shapes as well as correct natural frequencies, but it is unlikely that it could have the correct stiffness distribution and correct natural frequencies while having incorrect mode shapes.

References

1. Spillers WR, MacBain KM (2009) Structural optimization. Springer, Dordrecht
2. Wright JR, Cooper JE (2015) Introduction to aircraft aeroelasticity and loads, 2nd edn. Wiley, Chichester
3. Hodges DH, Pierce GA (2011) Introduction to structural dynamics and aeroelasticity. Cambridge University Press, New York
4. French M, Eastep FE (1996) Aeroelastic model design using parameter identification. J Aircr 33 (1):198–202
5. Caidin M (2012) Fork-tailed devil: the P-38. ibooks, New York

Chapter 10
Multiobjective Optimization

Life is just a series of trying to make up your mind

-Timothy Fuller

Don't give me what I asked for, give me what I wanted.

-Pretty much everyone

10.1   The Need for Multiobjective Optimization

It's not unusual for a design problem to have no clear single objective function. It is not hard to see how this might happen. Imagine the task of designing a low-cost passenger car – the task faced by Henry Ford in the early 1900s. There were many small car manufacturers, and the average car was too expensive for the average worker. He needed to make a car that was simultaneously affordable, reliable, easy to repair, adaptable, attractive enough to appeal to potential buyers, and a list of other things.

Figure 10.1 shows a 1912 Model T roadster (the photograph was taken in 1913). It had two rather than four seats and a large fuel tank behind the cabin. It is worth noting that this one, owned by Dr. (later Sir) David Hardie, is right-hand drive. Apparently, the car was adaptable enough that this modification didn't cause undue problems. That the Model T was in production from 1908 to 1927 says much about how well it served the needs of its market.

It's highly unlikely that Mr. Ford used formal optimization methods to design the Model T, but it is worth considering what might have been considered and what the effects might have been. It certainly would have been possible to treat cost as the objective function and other characteristics as constraints. However, this might have driven him to minimize production cost on each of the components, even if that meant reduced life or increased difficulty of maintenance.

Conversely, if durability was the objective function, he might have been drawn to a design in which all the parts were very heavily built. Even if that, somehow, didn't increase their cost, it might result in a car so heavy, that its fuel consumption suffered. It's perhaps interesting to note that the engine could run on gasoline, kerosene, or ethanol and that Ford advertised fuel economy on the order of

M. French, *Fundamentals of Optimization*,
https://doi.org/10.1007/978-3-319-76192-3_10

Fig. 10.1 A Model T Roadster 1913, driven by Dr. and Mrs. David Hardie. (Wikimedia Commons, image is in the public domain)

18-11L/100 km (13–21 mi/gal). Ford clearly intended his design to be attractive to people without unlimited resources.

The point is that there could certainly have been several choices of objective function and each choice would have conditioned the resulting design in a different way. However, the needs were so broad that it may not have been possible to choose a single objective function. A more desirable result might have come from choosing several objective functions, such as cost and durability. These two would seem to be naturally opposed to one another, and this is usually the nature of multiobjective problems. If they weren't working at cross-purposes, it might not really be a multiobjective problem.

Finally, the basic design of the Model T was so adaptable that, though intended to be utilitarian, it was the basis of racing cars, like the Speedster shown in Fig. 10.2, and several generations of "hot rods," some of which have engines producing hundreds of horsepower, a far cry from the 20 hp (15 kW) of the original engine.

10.2   A Simple Multiobjective Problem

To start, let's consider a very simple, unconstrained multiobjective problem.

Fig. 10.2   A Speedster racing car developed from the Model T. (Wikimedia Commons, image is in the public domain)

$$\text{Minimize } \begin{array}{l} f_1 = x^2 - 2x \\ f_2 = e^{-x} + e^{4x} \end{array} \qquad (10.1)$$

Figure 10.3 shows a plot of these two functions, including their minima.

10.3   Weighted Objectives Method

The most obvious method of accommodating more than one objective function is to simply sum them, with appropriate weights assigned [1, 2].

$$f(x) = \sum_{i=1}^{N} w_i f_i(x) \qquad (10.2)$$

where w_i are scalar weights. While easy to implement, the results can be hard to interpret since the values of the weighting functions aren't always clearly related to the relative importance objective functions. They are sometimes assigned empirically based on previous calculations.

As an example, let's assume $w_1 = 1$ and $w_2 = 2$. In this case, the combined objective function is

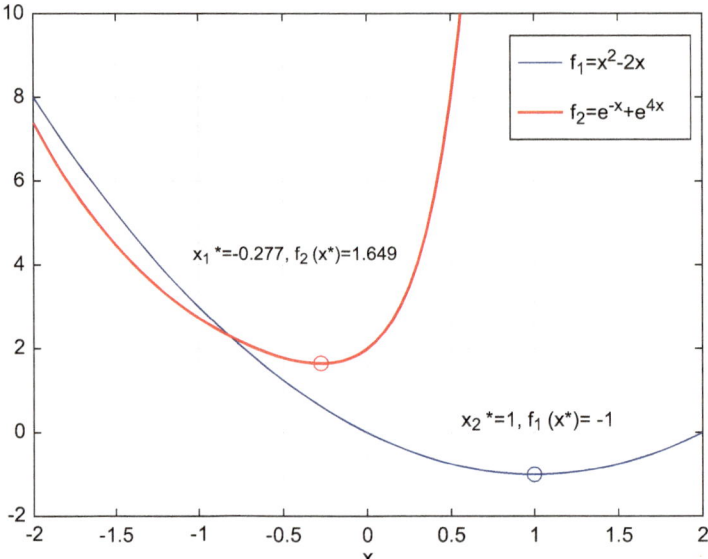

Fig. 10.3 Two-sample functions for a multiobjective problem

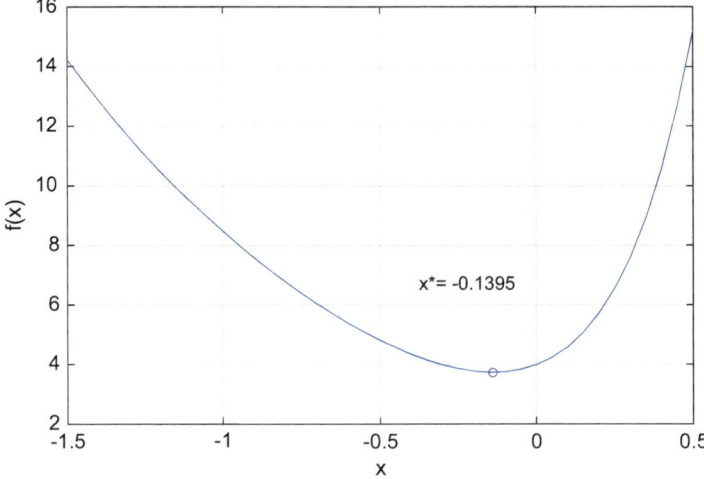

Fig. 10.4 Minimum of weighted objective functions

$$f(x) = x^2 - 2x + 2\left[e^{-x} + e^{4x}\right] \tag{10.3}$$

Figure 10.4 shows the result of minimizing the weighted objective function.

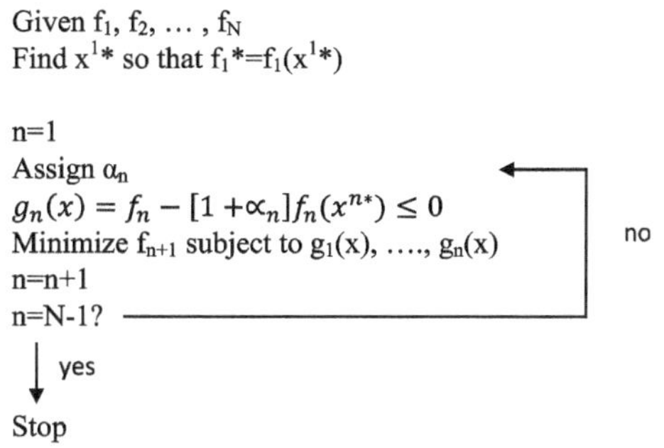

Fig. 10.5 Pseudo flow
chart for hierarchical
optimization method

Given $f_1, f_2, \ldots, f_N$
Find x^{1*} so that $f_1* = f_1(x^{1*})$

n=1
Assign α_n
$g_n(x) = f_n - [1 + \alpha_n] f_n(x^{n*}) \leq 0$
Minimize f_{n+1} subject to $g_1(x), \ldots, g_n(x)$ no
n=n+1
n=N-1? ————————————————

↓ yes

Stop

10.4 Hierarchical Optimization Method

Clearly, an improved method would be one that orders the objective functions by
their relative importance. In the hierarchical optimization method, each objective
function is minimized successively. A new constraint is added at each step – one for
each of the previously optimized objective functions. Each constraint essentially
limits the effect subsequent objective functions can have on the previous results, so it
is easier to reflect the relative importance of the individual objective functions.

The form of the additional constraints is

$$g_n(x) = f_n(x) - (1 + \alpha_n) f_n(x^{n*}) \leq 0 \qquad for \; n = 1, 2, \ldots N \qquad (10.4)$$

where α is a fractional change. This constraint might be interpreted as an effort to
limit the change in the multiobjective minimum from that of the first objective
function.

The pseudo flow chart is shown in Fig. 10.5. Note that x^{n*} is the location of the
minimum for the nth objective function.

Note that α shifts an objective function and uses the result as a constraint. For this
example, α must be negative. Given the two objective functions above and assuming
$\alpha = -1.5$, the solution proceeds as

Step 1: $\alpha = -1.5$

$x^* = 1$ and $f_1^* = -1$

Step 2: minimize f_2 subject to $g(x) = f_1(x) - (1 + \alpha)f_1^* \leq 0$

$x_m^* = -0.225 \qquad f_2(x^*) = f_m^* = 1.659$

It might be helpful to see what this looks like graphically. Figure 10.6 shows the
minima of both objective functions along with the multiobjective minimum when
$\alpha = -1.25$. The subscript m indicates the multiobjective minimum.

Finally, it is worth looking graphically at the effect of changes in α. Figure 10.7
shows the effect on the feasible region of increasing α to -0.5. The first objective
function is now more heavily weighted and x_m^* has been moved to the right.

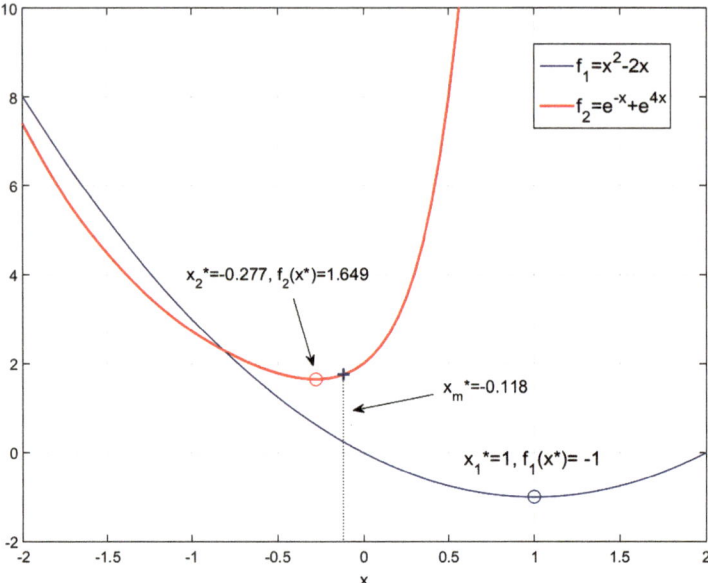

Fig. 10.6 Multiobjective minimum compared to individual minima

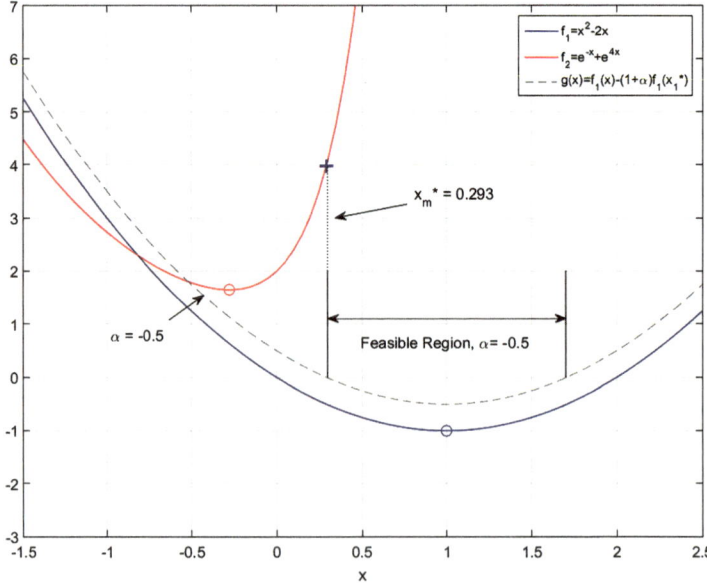

Fig. 10.7 Effect of increasing α

10.5   Global Criterion Method

Another simple method, global criterion [4], minimizes the distance between some
ideal reference point in design space and the feasible region. The multiobjective
problem is cast as a single objective function problem using a global criterion. The
objective function is

$$f(x) = \sum_{i=1}^{N} \left[\frac{f_i^0 - f_i(x)}{f_i^0} \right]^s \qquad (10.5)$$

If x^0 is the selected point in design space, then $f_1^0 = f_1(x^0)$, $f_2^0 = f_2(x^0)$, and so
on. Different values of s can be used, though the resulting minima are a strong
function of s. An alternate formulation casts the objective function as an L_p norm.

$$L_p(f) = \left[\sum_{i=1}^{N} \left| \frac{f_i^0 - f_i(x)}{f_i^0} \right|^p \right]^{1/p} \qquad 1 \le p \le \infty \qquad (10.6)$$

Note that, when $p = 2$, L_2 becomes a normalized Euclidian distance between the
value of the function and the ideal solution. It is not difficult to apply this method to
the test problem. If $p = 2$, the summed objective function is

$$L_2(f) = \left[\left[\frac{f_1^0 - f_1(x)}{f_1^0} \right]^2 + \left[\frac{f_2^0 - f_2(x)}{f_2^0} \right]^2 \right]^{1/2} \qquad (10.7)$$

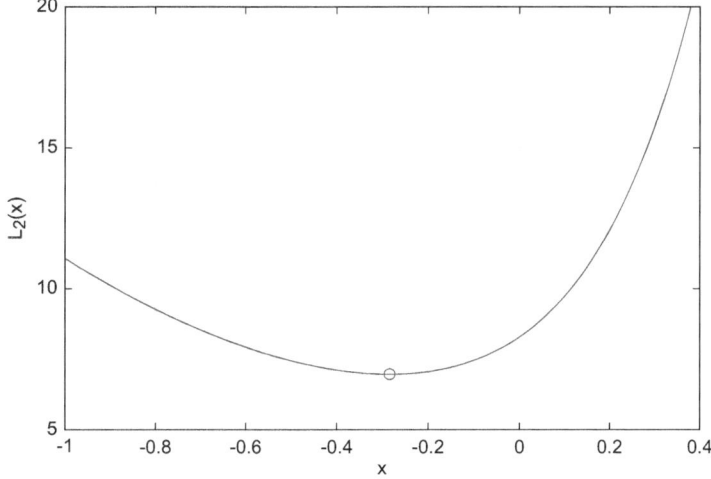

Fig. 10.8 Result from global criterion method

Using the values $f_1^* = 1$ and $f_2^* = -0.277$, L_2 is shown in Fig. 10.8. The minimum is $x^* = -0.283$ and $f^* = 6.964$.

10.6 Pareto Optimality

The previous sections have shown some basic methods for multiobjective optimization. With these in mind, it may be helpful to consider the big idea at work. When there is more than one objective function, there is a condition called Pareto optimality. It was named after Vilfredo Pareto, an engineer and economist who applied the concept to problems in economics [5]. He was a very colorful person who left, among the other elements of his academic legacy, a simple description of what an optimum means when there is more than one objective function.

A Pareto optimum is the location in design space such that any change in the values of the design variables increases the value of at least one objective function.

10.7 A Multiobjective Inverse Spectral Problem

Let's apply multiobjective methods to a problem with practical implications, an inverse spectral problem. Musical instruments, such as tubular bells, xylophones, and marimbas, use vibrating bars to make musical notes. Figure 10.9 shows a set of high-quality tubular bells manufactured by Yamaha. Here, the bars are hollow tubes but still vibrate normal to their axis, in bending.

The problem inherent with vibrating bars is that their resonant frequencies don't form a harmonic series. That is, their higher resonant frequencies are not generally integer multiples of the fundamental. This means that the higher resonant frequencies may not correspond to the frequency of any note in the needed key. Improvements in recording technology and theater acoustics mean that inharmonicity in musical instruments is increasingly unacceptable [6].

One approach is to cut away some of the structure in such a way that the resonant frequencies are closer to the notes in the needed scale. The bars used in marimbas (Fig. 10.10) are carved away for this reason.

Let's consider a thin bar that vibrates in bending, like a tubular bell. The easiest way to tune the frequency of a vibrating bar is to add masses to it. In the case of a tubular bell, they may take the form of collars fixed to the appropriate places on the tube. An analytical solution to this problem would be very difficult, so we'll use a finite element model. Figure 10.11 shows a simplified model with added masses. The model is symmetric about the centerline. We assume the bar is placed on soft supports so it has free-free boundary conditions.

Say the goal is to design a chime with a fundamental frequency of 110 Hz (note A_2 in the equal tempered scale) and a second frequency of 220 Hz – an octave higher. Eigenvalues are expressed in rad/s; the desired values are $\omega_1 = 2\pi \times 110 = 691.15$ and $\omega_2 = 2\pi \times 220 = 1382.3$.

Fig. 10.9 A set of tubular bells. (Wikimedia Commons, image is in the public domain)

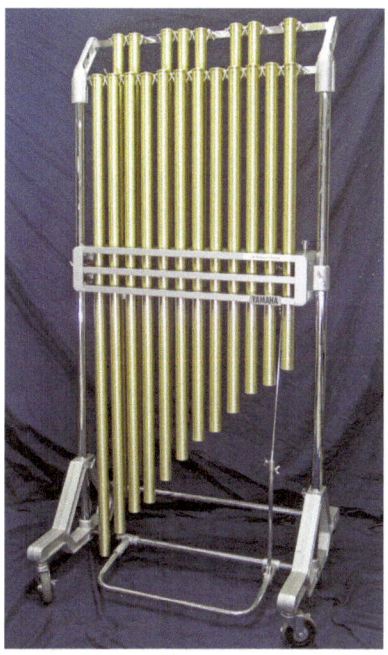

Fig. 10.10 Bars in a marimba from the Grinnell College Musical Instrument Collection. (omeka1. grinell.edu/MusicalInstruments)

We should note that, to keep the problem manageable, there are only four elements. In practice, more elements would be needed in order to correctly predict a useful number of natural frequencies. To verify that the underlying analysis is

Fig. 10.11 Simplified finite element model of a vibrating bar with masses

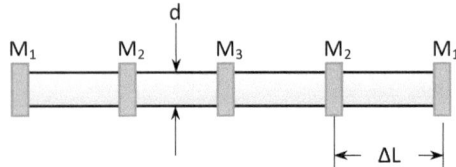

Table 10.1 Frequencies of plain bar

Eigenvalue number	Frequency (rad/s)
1	0
2	0
3	405.99
4	1124.9
5	2208.9

Table 10.2 Analytical and numerical frequency ratios

Ratio	Analytical	Numerical
ω_2/ω_1	2.76	2.77
ω_3/ω_1	5.40	5.44
ω_4/ω_1	8.93	9.97

correct, we can calculate the first three frequencies of a plain bar (no tuning masses) using realistic values of the model parameters.

Let's consider a round bar made of brass (UNS C2300, H55) so that $\rho = 8750$ kg/m^3 and $E = 115$ GPa. The overall length is 1 m and the diameter is 20 mm. Thus $r = 0.010$ m and $L_e = 0.250$ m. There are no tuning masses, so $m_1 = m_2 = m_3 = 0$.

The first five calculated frequencies are shown in Table 10.1.

Note that the first two frequencies are zero. They were actually very small – on the order of 10^{-5} – as is routine in numerical calculations. The two zero frequencies correspond to rigid body modes of the beam. Since it has free boundary conditions at both ends, there is nothing to prevent the whole beam from moving vertically or from rotating. The first flexible mode corresponds to the third eigenvalue.

It is routine in analytical beam calculations to express frequencies as ratios since they are more general; the properties of specific beams divide out. For a freely supported beam, the frequency ratios for the first few flexible modes are shown in Table 10.2. Differences between the analytical and numerical values will decrease as the number of elements in the numerical model increases.

As a first step in the multiobjective problem, consider a weighted summation whose terms are the unweighted, normalized squared frequency errors. Note that the chosen frequencies are 110 Hz and 220 Hz respectively. These correspond to the note A$_2$ in the equal tempered scale and its first harmonic.

$$F = \left[\frac{\omega_1 - 691.15}{691.15}\right]^2 + \left[\frac{\omega_2 - 1382.3}{1382.3}\right]^2 \tag{10.8}$$

Table 10.3 Optimal design using weighted summation

Parameter	Value
r	0.01467 m
L_e	0.2251 m
m_1	0.04961 kg
m_2	3.7179 kg
m_3	0.09259 kg

Table 10.4 Optimal design using global criterion

Parameter	Value
r	0.02494 m
L_e	0.27351 m
m_1	1.020 kg
m_2	11.59 kg
m_3	0 kg

Table 10.5 Preferred design using global criterion

Parameter	Value
r	0.0060583 m
L_e	0.14827 m
m_1	0 kg
m_2	0.4101 kg
m_3	0 kg

The resulting design is presented in Table 10.3. The design almost exactly minimizes the two error functions, so that $\omega_1 = 691.126$ rad/s and $\omega_2 = 1382.35$ rad/s.

Note that this is just one of many possible designs. A number of different starting points were evaluated, and it appears, as with most practical problems, that there are many local minima.

It is a simple extension to move from the summation to the global criterion method.

$$F = \left[\left[\frac{\omega_1 - 691.15}{691.15} \right]^2 + \left[\frac{\omega_2 - 1382.3}{1382.3} \right]^2 \right]^{1/2} \tag{10.9}$$

The initial result is a much different design that is equally acceptable as shown in Table 10.4. Again, the design almost exactly minimizes the two error functions, so that $\omega_1 = 690.831$ rad/s and $\omega_2 = 1382.34$ rad/s.

Note that the second mass would be very heavy. While this design is mathematically acceptable, it is not necessarily practical.

Experimenting with different starting points gives a design that is not only practical but also satisfies the frequency requirements to several decimal places. Table 10.5 shows the results for a more practical design in which $\omega_1 = 691.15$ rad/s and $\omega_2 = 1382.3$ rad/s.

Fig. 10.12   A Mark 48 torpedo (Wikimedia Commons, image is in the public domain)

These predictions are, of course, accurate only to the limits of the underlying analysis but are encouraging given the difficulty of designing chimes to improve sound quality.

10.8   A Complex Multiobjective Design Problem

As described above, practical design problems can be complex enough that identifying a single objective function is quite difficult. As one possible example, consider the design of a torpedo. It is the underwater analog of a guided missile and has many conflicting design requirements, so many that there may not be a clear objective function.

Its job is fairly simple in concept. It rides around in a submarine, perhaps for a very long time, until it is fired at some target, such as a ship. It must be guided or guide itself to the target ship and explode with enough force to incapacitate it. Figure 10.12 shows the Mark 48 torpedo, the design currently used by the US Navy.

From the outside, it looks like a cylinder with a partially streamlined nose and a shrouded propeller at the back – pretty simple. However, this belies the complexity that lies within. An almost assuredly incomplete list of requirements includes:

- Neutral buoyancy – It can't sink to the bottom or float to the surface upon firing.
- Long range – Getting too close to one's target before firing is usually unwise.
- Destructive power – It needs to be able to sink its target.
- Quiet – Too much acoustic warning gives the target time to react.
- Reliable – It may sit unused for quite some time before being fired and it has to work.
- Safe – It can't be a threat to its boat or the crew.
- Accurate – Can be remotely guided or self-guided but can't miss; once you've fired, you've announced both your presence and your intentions.
- Compatible – Must fit into a standard US Navy 21 inch tube.
- Discrete – Cannot leave a wake that would give away firing location.
- Depth – Must be able to withstand water pressure at extreme depths – "at least 1200 ft" or 366 m, according to public sources.

The details of operation are far beyond this brief example (and certainly closely held secrets), but it is enough to say that the Mark 48 torpedo is essentially an armed underwater drone. It can be fired under continuous control of the submarine through a trailing wire that plays out behind the torpedo. It can, for example, cruise after launch slowly and quietly for some distance from the submarine, perhaps on an indirect path, then turned toward the target. It might then be directed to accelerate to a high speed and seek the target using onboard sonar. The trailing wire can be cut whenever needed so that the firing tube may be reloaded. The targeted ship may, for example, be unaware of approaching torpedoes until the noise made by their propellers and their active sonar is detected, approaching the ship at high speed and from two widely spaced directions.

Now consider the design problem. Given the complex nature of the device, what should the objective function be? It is very hard to define a single design goal, so clearly, this is a possible application of multiobjective methods.

References

1. Gass S, Saaty T (1955) The computational algorithm for the parametric objective function. Naval Res Log Q 2:39–45
2. Branke J, Deb K, Miettinen K, Slowinski R (eds) (2008) Multiobjective optimization – interactive and evolutionary approaches. Springer, Berlin
3. Sartini de Oliveira L, Saramago SFP Multiobjective optimization techniques applied to engineering problems. J Braz Soc Mech Sci Eng 32(1):94–105
4. Zeleny M (1973) Compromise programming. In: Cochrane JL, Zeleny M (eds) Multiple criteria decision making. University of South Carolina Press, Columbia, pp 262–301
5. Amoroso L (1938) Vilfredo Pareto. Econometrica 6(1):1
6. Huston DH Personal Conversation, October 2017

Appendix A: Derivatives

Optimization relies heavily on slopes of functions called derivatives. It's worth reviewing where these derivatives come from and how they are extended to multiple independent variables.

Derivatives of a Single Variable

The first thing to remember is that a derivative is just a slope and the definition looks very much like the expression for slopes you learned in Jr. high school. Fig. A.1 shows slope defined using two points on a curve.

The slope of the line connecting the two points is

$$m = \frac{f(x + \Delta x) - f(x)}{\Delta x} \tag{A.1}$$

The only problem with this picture is that the two points are so far apart that the slope of the line is not necessarily close to the slope of the curve at either point. The obvious way to address this problem is to move the points closer together as shown in Fig. A.2. The slope of the line is clearly closer to the slope at the two points.

For the slope of the line to closely match the slope at the points, those points must be very close together. The distance between them cannot be zero because the expression for slope requires dividing by Δx. However, the two points can be infinitesimally close – as close as they need to be without the distance between them actually being zero. In mathematical notation, this is written as

$$\frac{df}{dx} = \lim_{\Delta x \to 0} \frac{f(x + \Delta x) - f(x)}{\Delta x} \tag{A.2}$$

© Springer International Publishing AG, part of Springer Nature 2018
M. French, *Fundamentals of Optimization*,
https://doi.org/10.1007/978-3-319-76192-3

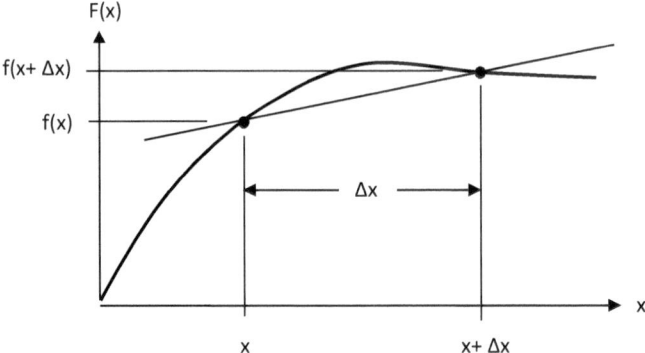

Fig. A.1 Slope defined using two points

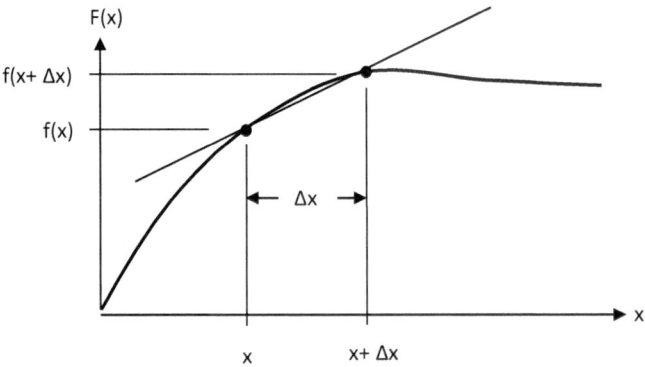

Fig. A.2 Points moved closer together

This is the definition of the derivative. To make this more concrete, let's assume that $f(x) = x^2$ and work through the derivative expression. Simply substituting this function into the expression for the derivative gives

$$\frac{df}{dx} = \lim_{\Delta x \to 0} \frac{(x + \Delta x)^2 - x^2}{\Delta x} \tag{A.3}$$

Expanding the expression out gives

$$\frac{df}{dx} = \lim_{\Delta x \to 0} \frac{x^2 + 2x\Delta x + \Delta x^2 - x^2}{\Delta x} \tag{A.4}$$

Which simplifies to

$$\frac{df}{dx} = \lim_{\Delta x \to 0} 2x + \Delta x \qquad (A.5)$$

Since Δx is arbitrarily small, $df/dx = 2x$, just as we expect.

Clearly we don't want go through this limit process every time we need to find a derivative. Fortunately, there are tables of derivatives for commonly used expressions. Also, there are rules for finding derivatives of complicated functions. The two most commonly used are probably the chain rule and the product rule.

Derivatives of More than One Variable

Math is, by its most fundamental nature, general. Thus, a mathematical operation defined for one kind of expression is often defined for many others as well. The simplest example might be addition. Integers can be added together. Similarly, so can fractions, real numbers, complex numbers, functions, vectors, matrices, etc. Sometimes the details change, for example, not every matrix can be added to every other matrix. However, the basic idea holds.

Similarly, derivatives are defined for functions of more than one variable. This more general form of derivative is called a gradient, and it is central to many optimization methods. The way to identify a gradient when reading an equation is to note that "d" is replaced by ∂, called a "partial." Thus, the expression $\partial y/\partial x$ is called "partial y, partial x."

The symbol ∂ indicates that the expression has more than one independent variable. A simple function of more than one variable is

$$f(x, y) = x^2 + y^2 + \cos(2xy) + x^2 y^2 \qquad (A.6)$$

The derivative with respect to x, $\partial f/\partial x$, is the derivative of f, taken assuming that x is a variable and y is a constant. Similarly, the derivative with respect to y, $\partial f/\partial y$, is the derivative of f, taken assuming that y is a variable and x is a constant.

$$\nabla f(x, y) = \left\{ \begin{array}{c} \dfrac{\partial f}{\partial x} \\[2mm] \dfrac{\partial f}{\partial y} \end{array} \right\} = \left\{ \begin{array}{c} 2x + 2xy^2 - 2y\,\sin(2xy) \\[2mm] 2y + 2x^2 y - 2x\,\sin(2xy) \end{array} \right\} \qquad (A.7)$$

The result is a vector with direction and magnitude. The gradient has a property that makes it especially useful in optimization. The direction of the gradient vector is the direction of steepest slope at a point. Remember that, for a function, the slope depends both on where you are and what direction you are looking. At a given location, you could turn to look uphill (maximum slope) to the side (zero slope) or downhill (minimum slope). Note that the direction of minimum slope is just the negative of the direction of maximum. Thus, the direction of minimum slope, S_{min}, is $S_{min} = -\nabla f$.

Appendix B: Test Functions

When learning optimization methods, it helps to have a collection of test functions. Here is a collection of test functions along with their minima and derivatives.

Single Variable Functions

Function 1

$$f_1(x) := x^4 - 100 \cdot x^2 - 10 \cdot x + 1$$

$$df_1(x) := \frac{d}{dx}\left(x^4 - 100 \cdot x^2 - 10 \cdot x + 1\right) \rightarrow 4 \cdot x^3 - 200 \cdot x - 10$$

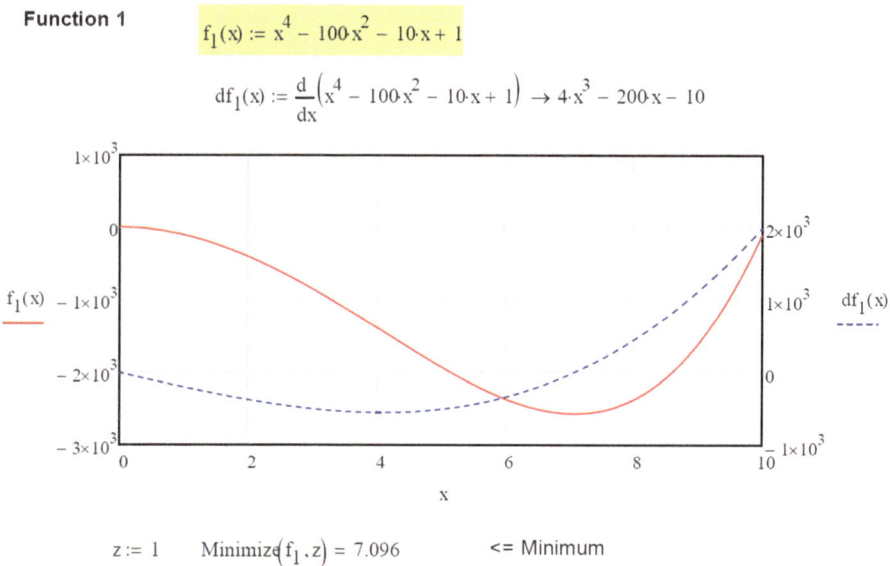

$$z := 1 \qquad \text{Minimize}\left(f_1, z\right) = 7.096 \qquad <= \text{Minimum}$$

Fig. B.1 Single variable test function 1

© Springer International Publishing AG, part of Springer Nature 2018
M. French, *Fundamentals of Optimization*,
https://doi.org/10.1007/978-3-319-76192-3

Function 2

$$f_2(x) := -x \cdot e^{-\frac{x}{5}} \qquad df_2(x) := \left(-e^{-\frac{1}{5} \cdot x}\right) + \frac{1}{5} \cdot x \cdot e^{-\frac{1}{5} \cdot x}$$

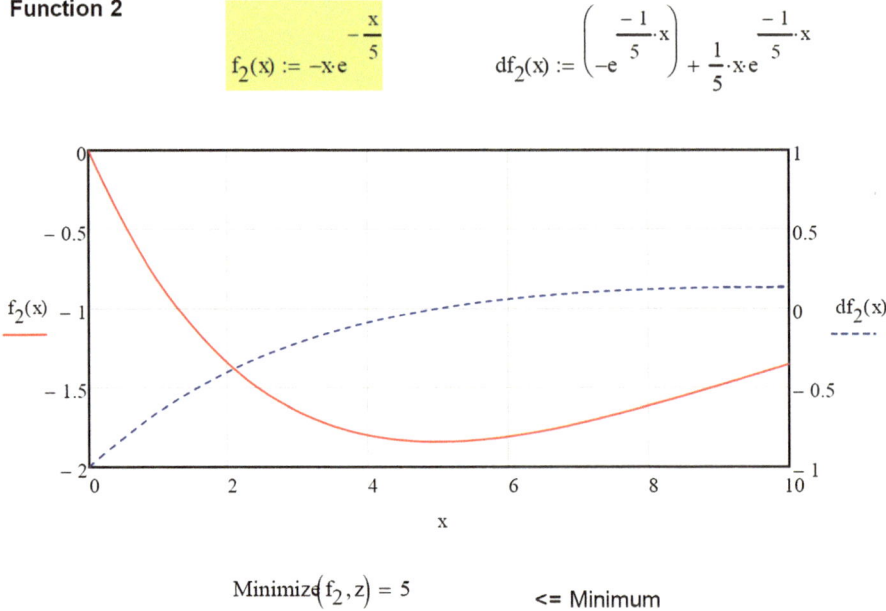

$$\text{Minimize}(f_2, z) = 5 \qquad\qquad \texttt{<= Minimum}$$

Fig. B.2 Single variable test function 2

Function 3

$$f_3(x) := \frac{-x^3 - x^2 + 1}{x^3 - x^2 + 1} \qquad df_3(x) := \frac{(-3) \cdot x^2 - 2 \cdot x}{x^3 - x^2 + 1} - \frac{\left(-x^3\right) - x^2 + 1}{\left(x^3 - x^2 + 1\right)^2} \cdot \left(3 \cdot x^2 - 2 \cdot x\right)$$

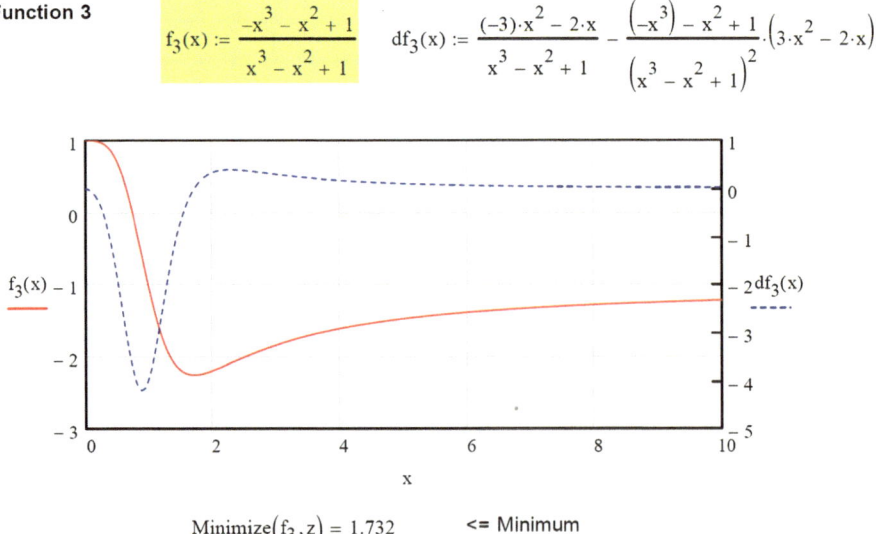

$$\text{Minimize}(f_3, z) = 1.732 \qquad \texttt{<= Minimum}$$

Fig. B.3 Single variable test function 3

Function 4

$$f_4(x) := \sin\left(\frac{3}{2}x\right) \cdot x \cdot e^{\frac{-x}{2}} + \frac{x^2}{100}$$

$$df_4(x) := \frac{3}{2} \cdot \cos\left(\frac{3}{2} \cdot x\right) \cdot x \cdot e^{\frac{-1}{2} \cdot x} + \sin\left(\frac{3}{2} \cdot x\right) \cdot e^{\frac{-1}{2} \cdot x} - \frac{1}{2} \cdot \sin\left(\frac{3}{2} \cdot x\right) \cdot x \cdot e^{\frac{-1}{2} \cdot x} + \frac{1}{50} \cdot x$$

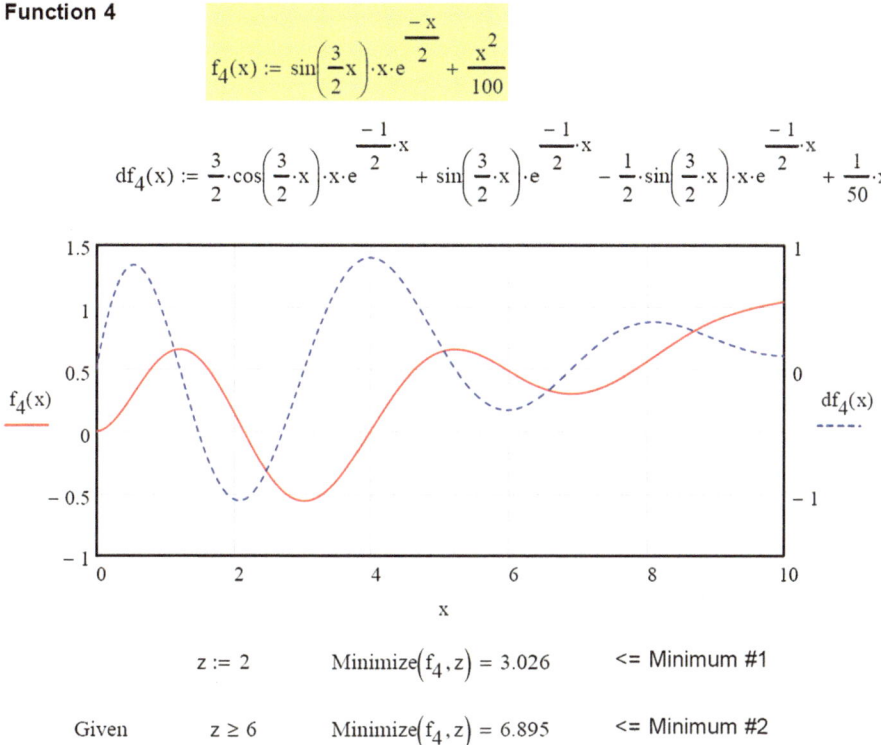

$$z := 2 \qquad \text{Minimize}(f_4, z) = 3.026 \qquad \text{<= Minimum \#1}$$

$$\text{Given} \qquad z \geq 6 \qquad \text{Minimize}(f_4, z) = 6.895 \qquad \text{<= Minimum \#2}$$

Fig. B.4 Single variable test function 4

Function 5 $\qquad f_5(x) := (x - 1) \cdot (x - 2) \cdot (x - 10) \cdot (x - 7) \text{ expand } \rightarrow x^4 - 20 \cdot x^3 + 123 \cdot x^2 - 244 \cdot x + 140$

$$df_5(x) := 4 \cdot x^3 - 60 \cdot x^2 + 246 \cdot x - 244$$

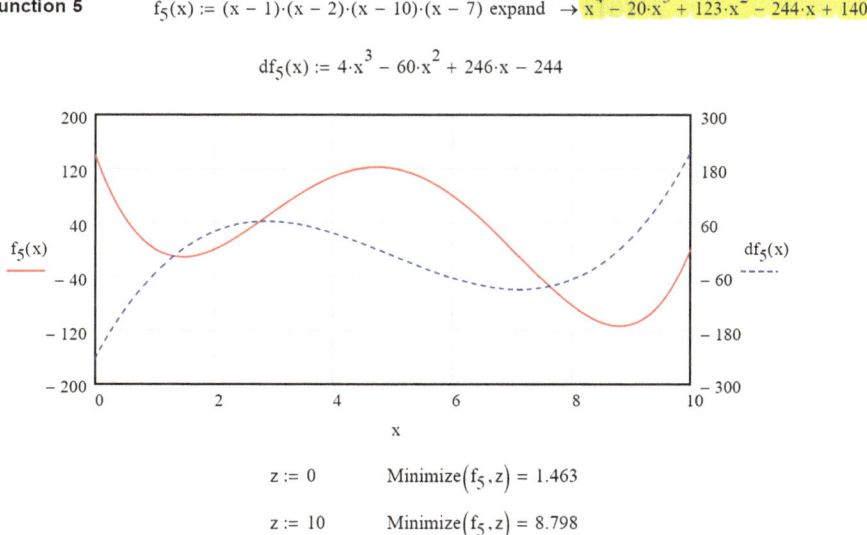

$$z := 0 \qquad \text{Minimize}(f_5, z) = 1.463$$

$$z := 10 \qquad \text{Minimize}(f_5, z) = 8.798$$

Fig. B.5 Single variable test function 5

Function 6

$$f_6(x) := \sin(2 \cdot x) \cdot e^{-(0.4 \cdot x)^2} + \frac{(x-1)^2 + 1}{1000}$$

$$x := -10, -9.9 .. 10$$

$$df_6(x) := 2 \cdot \cos(2 \cdot x) \cdot e^{(-.16) \cdot x^2} - .32 \cdot \sin(2 \cdot x) \cdot x \cdot e^{(-.16) \cdot x^2} + \frac{1}{500} \cdot x - \frac{1}{500}$$

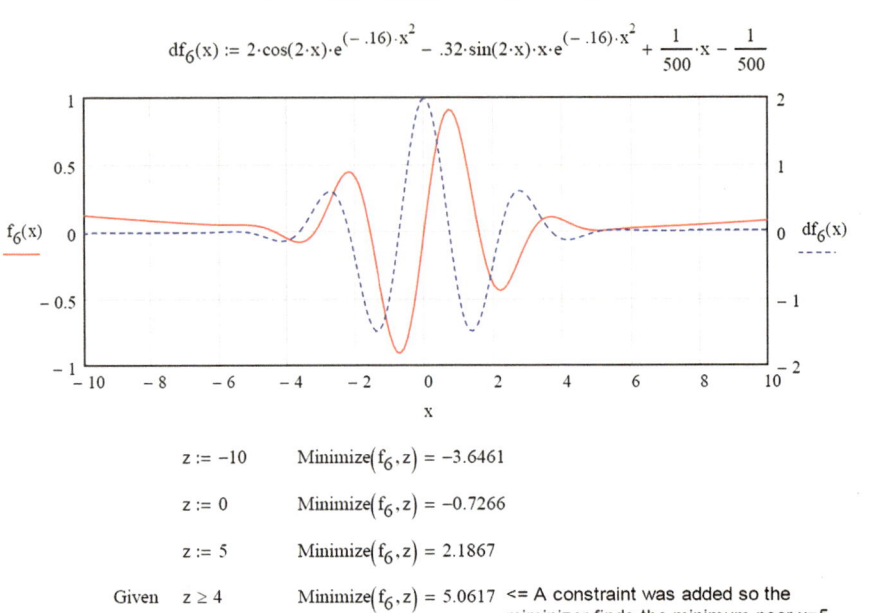

$z := -10$	$\text{Minimize}(f_6, z) = -3.6461$
$z := 0$	$\text{Minimize}(f_6, z) = -0.7266$
$z := 5$	$\text{Minimize}(f_6, z) = 2.1867$
Given $z \geq 4$	$\text{Minimize}(f_6, z) = 5.0617$ <= A constraint was added so the miminizer finds the minimum near x=5

Fig. B.6 Single variable test function 6

Function 7

$$f_7(x) := \ln\left[\left(e^{-x} + 1\right) \cdot \left(e^{\frac{x}{2}} + 1\right)\right]$$

$$df_7(x) := \frac{-1}{2} \cdot \frac{e^{\frac{1}{2} \cdot x} + 2 - e^{\frac{3}{2} \cdot x}}{\left(1 + e^x\right) \cdot \left(e^{\frac{1}{2} \cdot x} + 1\right)}$$

$$\text{Minimize}(f_7, z) = 0.8392$$

Fig. B.7 Single variable test function 7

Two Variable Functions

Rosenbrock Banana Function

$$b(x,y) := (1 - x)^2 + \left(y - x^2\right)^2$$

$$dbdx(x,y) := \frac{d}{dx}b(x,y) \rightarrow 2 \cdot x - 4 \cdot x \cdot \left(y - x^2\right) - 2$$

$$dbdy(x,y) := \frac{d}{dy}b(x,y) \rightarrow 2 \cdot y - 2 \cdot x^2$$

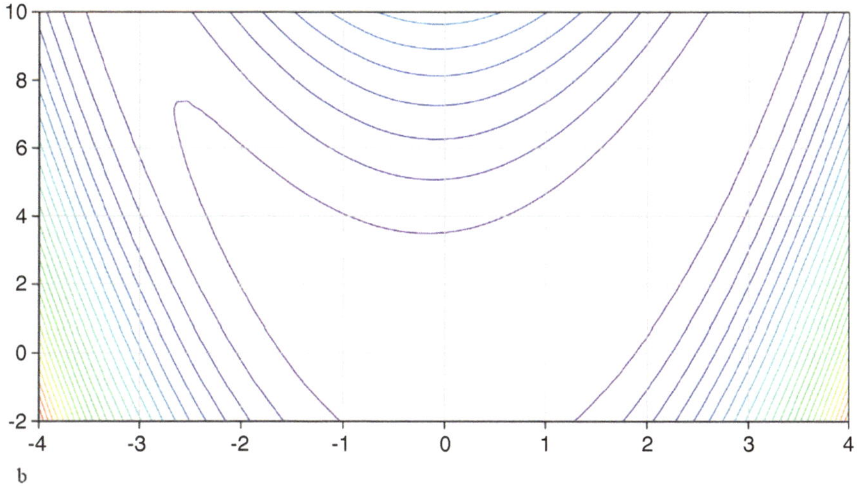

b

$x := 2 \qquad y := 0 \qquad \text{Minimize}(b,x,y) = \begin{pmatrix} 1 \\ 1 \end{pmatrix}$ <= Minimize using built-in function

$dbdx(1,1) = 0$ <= Check that gradient is zero at minimum

$dbdy(1,1) = 0$

Fig. B.8 Two variable test function 1, Rosenbrock banana function

Butterfly Function

$$f(x,y) := \frac{x - y}{\left(x^2 + 5\right) \cdot \left(y^2 + 5\right)} + \frac{y^2}{20000}$$

$$\frac{d}{dx} f(x,y) \rightarrow \frac{1}{\left(x^2 + 5\right) \cdot \left(y^2 + 5\right)} - \frac{2 \cdot x \cdot (x - y)}{\left(x^2 + 5\right)^2 \cdot \left(y^2 + 5\right)}$$

$$\frac{d}{dy} f(x,y) \rightarrow \frac{y}{10000} - \frac{1}{\left(x^2 + 5\right) \cdot \left(y^2 + 5\right)} - \frac{2 \cdot y \cdot (x - y)}{\left(x^2 + 5\right) \cdot \left(y^2 + 5\right)^2}$$

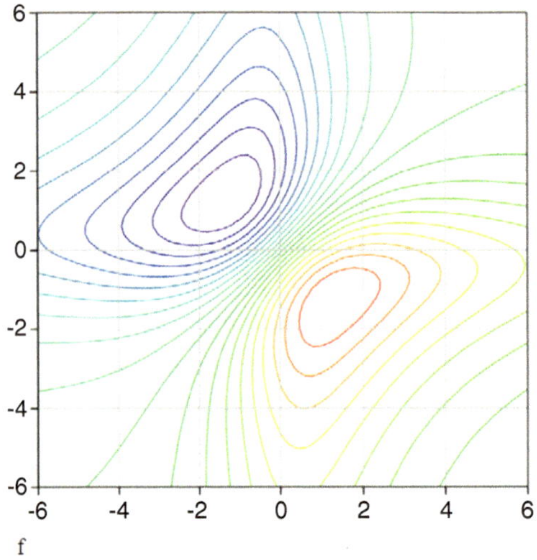

x := 1   <= Initial guesses for minimizer

y := 1

$$\text{Minimize}(f, x, y) = \begin{pmatrix} -1.296 \\ 1.281 \end{pmatrix}$$

f

Fig. B.9 Two variable test function 2, butterfly function

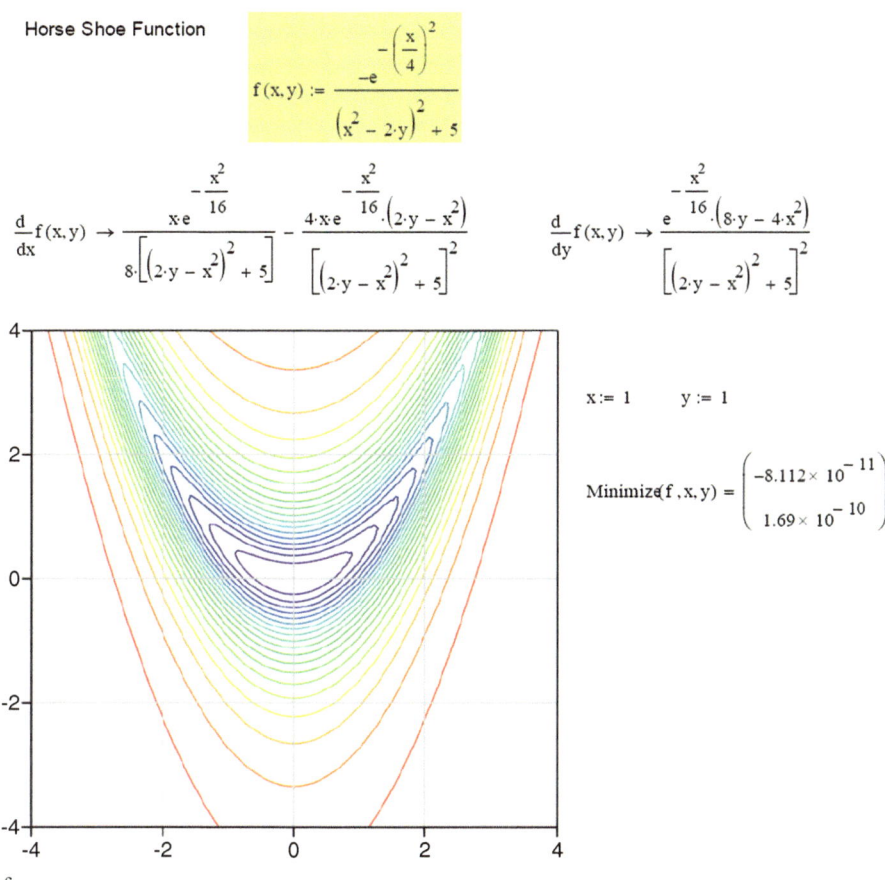

Fig. B.10 Two variable test function 3, horseshoe function

Zig Zag Function

$$f(x,y) := \dfrac{-e^{-\left(\frac{x}{2}\right)^2}}{\left(x^3 + x^2 - 2 \cdot y\right)^2 + 3}$$

$$\frac{d}{dx}f(x,y) \rightarrow \frac{x \cdot e^{-\frac{x^2}{4}}}{2 \cdot \left[\left(x^3 + x^2 - 2 \cdot y\right)^2 + 3\right]} + \frac{2 \cdot e^{-\frac{x^2}{4}} \cdot \left(3 \cdot x^2 + 2 \cdot x\right) \cdot \left(x^3 + x^2 - 2 \cdot y\right)}{\left[\left(x^3 + x^2 - 2 \cdot y\right)^2 + 3\right]^2}$$

$$\frac{d}{dy}f(x,y) \rightarrow \frac{e^{-\frac{x^2}{4}} \cdot \left(4 \cdot x^3 + 4 \cdot x^2 - 8 \cdot y\right)}{\left[\left(x^3 + x^2 - 2 \cdot y\right)^2 + 3\right]^2}$$

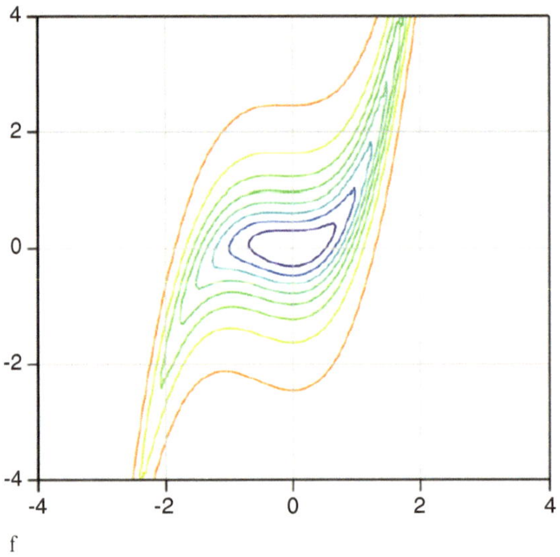

$$x := 1 \qquad y := 1$$

$$\text{Minimize}(f,x,y) = \begin{pmatrix} -4.377 \times 10^{-7} \\ 5.467 \times 10^{-9} \end{pmatrix}$$

f

Fig. B.11 Two variable test function 4, zigzag function

Splatter Function

$$f(x,y) := \log\left[\left(x^2 \cdot \cos(y) + y^2 \cdot \sin(x)\right)^2 + x^2 \cdot y^2 + (x+y)^2 + 1\right]$$

$$\frac{d}{dx}f(x,y) \rightarrow \frac{2 \cdot x + 2 \cdot y + 2 \cdot x \cdot y^2 + 2 \cdot \left(y^2 \cdot \cos(x) + 2 \cdot x \cdot \cos(y)\right) \cdot \left(x^2 \cdot \cos(y) + y^2 \cdot \sin(x)\right)}{\ln(10) \cdot \left[\left(x^2 \cdot \cos(y) + y^2 \cdot \sin(x)\right)^2 + (x+y)^2 + x^2 \cdot y^2 + 1\right]}$$

$$\frac{d}{dy}f(x,y) \rightarrow \frac{2 \cdot x + 2 \cdot y + 2 \cdot x^2 \cdot y - 2 \cdot \left(x^2 \cdot \sin(y) - 2 \cdot y \cdot \sin(x)\right) \cdot \left(x^2 \cdot \cos(y) + y^2 \cdot \sin(x)\right)}{\ln(10) \cdot \left[\left(x^2 \cdot \cos(y) + y^2 \cdot \sin(x)\right)^2 + (x+y)^2 + x^2 \cdot y^2 + 1\right]}$$

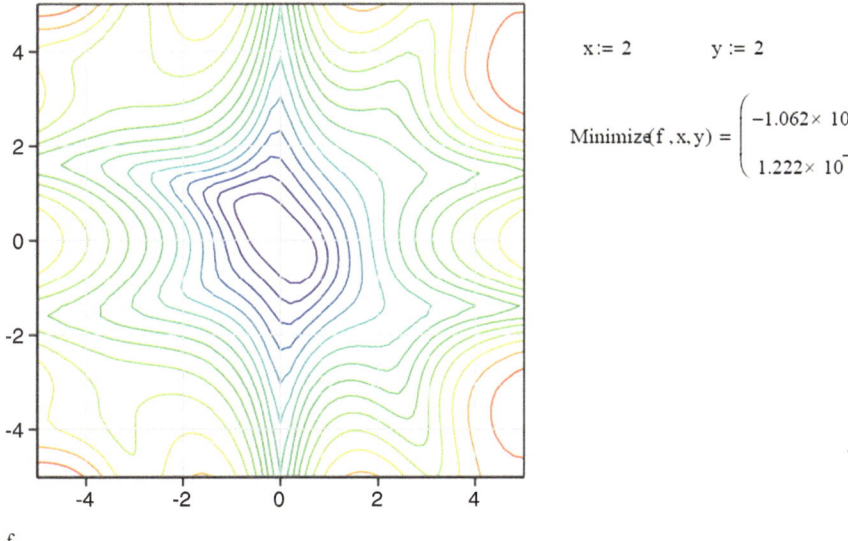

$$x := 2 \qquad y := 2$$

$$\text{Minimize}(f,x,y) = \begin{pmatrix} -1.062 \times 10^{-4} \\ 1.222 \times 10^{-4} \end{pmatrix}$$

f

Fig. B.12 Two variable test function 5, bug splatter function

Three Variable Function

Three Variable Test Function

$$f(x,y,z) := 2 \cdot x^2 + 2 \cdot y^2 + z^4 + 8 \cdot x + 4 \cdot y + 10 \cdot z + 2 \cdot e^{-\left(\frac{z}{4}\right)^2}$$

$$\frac{d}{dx} f(x,y,z) \rightarrow 4 \cdot x + 8 \qquad \frac{d}{dy} f(x,y,z) \rightarrow 4 \cdot y + 4 \qquad \frac{d}{dz} f(x,y,z) \rightarrow 4 \cdot z^3 - \frac{z \cdot e^{-\frac{z^2}{16}}}{4} + 10$$

$$x := -4 \qquad y := 8 \qquad z := 10 \qquad\qquad \text{Minimize}(f,x,y,z) = \begin{pmatrix} -2 \\ -1 \\ -1.371 \end{pmatrix}$$

Fig. B.13 Three variable test function

Appendix C: Solving Optimization Problems Using MATLAB

Clearly, meaningful optimization problems must be solved using numerical methods. At this writing, there are several different general purpose technical computation packages in widespread use, along with more specialized optimization software. However, the most popular is MATLAB, a technical programming language that is as close to a standard as there is.

In addition to the core language functions, there are toolboxes (collections of related functions) for many specialized uses. Some of these are distributed through MathWorks, the publisher of MATLAB, and some are supplied by others. The other suppliers are sometimes university research groups.

There is an optimization tool box provided MathWorks that contains a broad range of well-developed optimization functions. Since this book is intended as an introduction, we will focus on five simple, general purpose functions:

- fzero – Find the root of a nonlinear function
- fsolve – Solve systems of nonlinear equations
- fminsearch – Find unconstrained minimum of a nonlinear function without derivatives
- fminunc – Find unconstrained minimum of a nonlinear function
- fmincon – Find constrained minimum of a nonlinear function

Note that these examples were created using MATLAB R2015a and R2016b. MATLAB is generally good about being backward compatible, but there are occasional, inexplicable, small changes that can cause syntax problems. In particular, the options definition for fminunc seems to have changed slightly between the two versions of MATLAB available to the author. Notes appear as comments in the M files where applicable.

Finally, the capabilities of MATLAB are extremely broad, and there are many functions related to optimization. The small collection presented here is just a starting point.

© Springer International Publishing AG, part of Springer Nature 2018   229
M. French, *Fundamentals of Optimization*,
https://doi.org/10.1007/978-3-319-76192-3

Finding Roots with fzero

"fzero" is a very simple function that finds the roots of a single nonlinear function and doesn't require a derivative. To start, let's find the root of test Function 3 from Appendix B

$$f(x) = \frac{-x^3 - x^2 + 1}{x^3 - x^2 + 1} \tag{C.1}$$

There are several different ways to program the solution in MATLAB. One of these is

```
function c1_fzero
%
%  This M file finds the root of Test Function 3 using fzero
%

f=@(x) (-x^3-x^2+1)/(x^3-x^2+1) % Use anonymous function to define f(x)

ezplot(f,[0 4 -3 1]); % Plot function
hold on   % Hold so marker can be added to show root
grid on
axis square

x0=1; % Define initial guess
[x_root, f_root]=fzero(f,x0) % Find root using fzero
plot(x_root,f_root,'bo') % Show location of root on plot

text(x_root+0.5,f_root,['x_{root} =   ' num2str(x_root)])

return
```

The resulting plot is shown in Fig. C.1.

It is important to note that the behavior of the algorithm is affected by the choice of starting points. In this case, a starting point of $x = 1$ found the root with no problem. However, starting at $x = 0$ failed, returning $x_{root} = -0.7549$ and $f(x_{root}) = 5.5350 \times 10^{14}$. This algorithm doesn't use derivatives. Rather, it uses a combination of bisection, secant, and inverse quadratic interpolation methods.

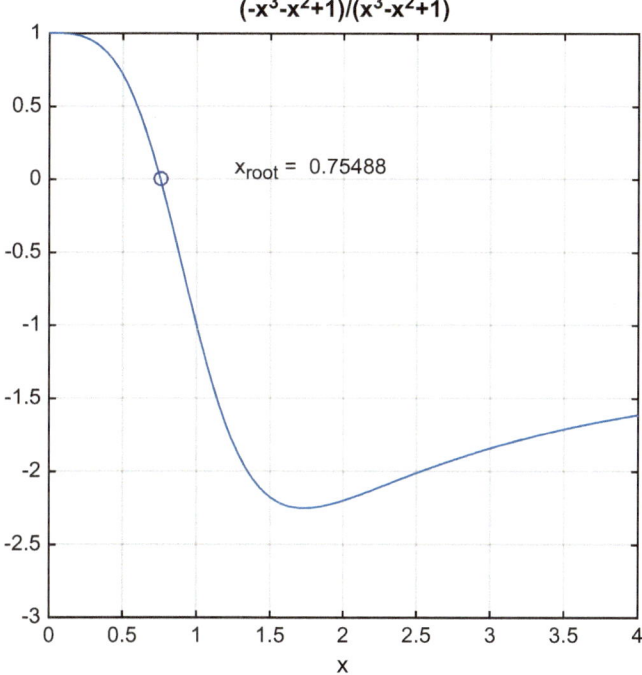

Fig. C.1 Finding root of test Function 3 using "fzero"

Another way to find the minimum of this function is to find the root of its derivative. There are two ways of identifying the derivative within MATLAB. The first is to simply hard code the derivative function.

$$f'(x) = \frac{-3x^2 - 2x}{x^3 - x^2 + 1} - \frac{-x^3 - x^2 + 1}{(x^3 - x^2 + 1)^2} (3x^2 - 2x) \qquad (C.2)$$

The other is to use symbolic capabilities in MATLAB as shown in this example

```
function c1_fzero_min
%
%   This M file finds the minimum of Test Function 3 using fzero
%
clc   % clear screen to keep things tidy
syms x % define x as a symbolic variable

f= (-x^3-x^2+1)/(x^3-x^2+1) % Define objective function
df=diff(f)   % Have MATLAB find derivative symbolically
ezplot(f,[0 4 -5 1]); % Plot function and derivative
hold on   % Hold so marker can be added to show minimum
ezplot(df,[0 4 -5 1]); % Plot function and derivative
grid on

% The following calculations are numerical rather than symbolic

fn=matlabFunction(f) % Make numerical function of symbolic function
dfdx=matlabFunction(df) % Make numerical function of derivative

x0=1; % Define initial guess
[x_min]=fzero(dfdx,x0) % Find root using fzero

plot(x_min,fn(x_min),'bo') % Show location of root on plot
text(x_min,fn(x_min)-0.5,['x_{min} =   ' num2str(x_min)])

hold off
return
```

The "syms x" command defines x as a symbolic variable so that $f(x)$ becomes a symbolic function. The command "diff(x)" tells MATLAB to find the symbolic derivative of the objective function. The resulting plot is shown in Fig. C.2.

Solving Systems of Equations Using Fsolve

From the user's point of view, fsolve can be considered as an extension of fzero because it works on systems of equations. Note that the two functions use different solution methods and fsolve can also make use of derivatives. Still, the command lines are similar.

To start, consider a very simple pair of coupled algebraic equations

$$
\begin{aligned}
y &= \cos(x) \\
y &= x^2
\end{aligned} \tag{C.3}
$$

The two functions are shown in Fig. C.3.

It is not difficult to set the two equations equal to one another to get $\cos(x) = x^2$ and to find the root. However, you choose to carry out the calculation; the solutions to this set of equations are $x = \pm 0.824$ and $y = 0.679$.

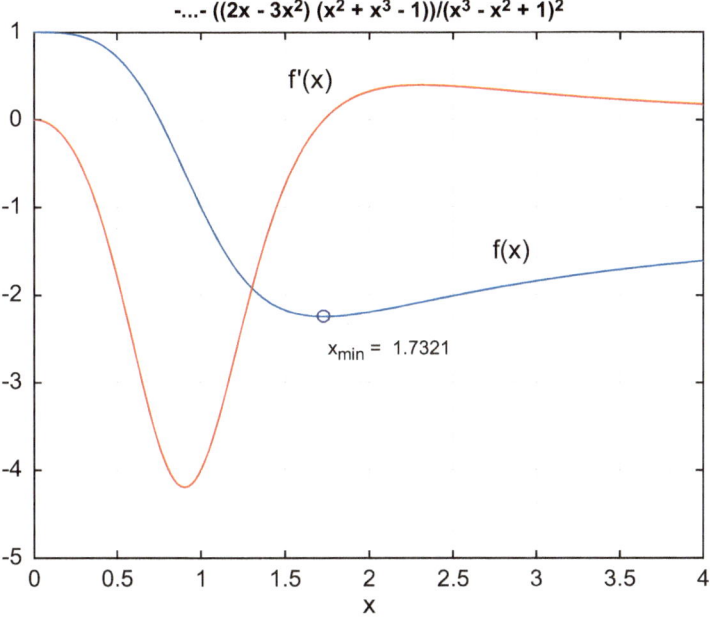

Fig. C.2 Finding minimum of test Function 3 using "fzero"

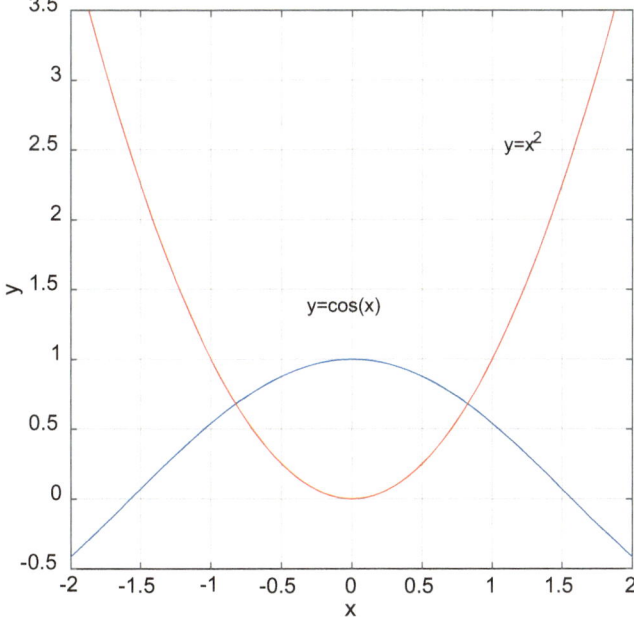

Fig. C.3 Test equations for "fsolve" example

Solving this system of equations using fsolve is mostly a matter of syntax. MATLAB assumes the equations are of the form $F(x) = 0$, so the test functions must be written as

$$y - \cos(x) = 0$$
$$y - x^2 = 0 \tag{C.4}$$

Here is a MATLAB example showing how to solve this problem using fsolve and defining the system of equations in a separate function. This would be akin to a subroutine in a more general programming language.

```
function ex1_fsolve
%
% Demonstrate fsolve using a function for system of equations
%
clc
fn=@eq2; % Define an anonymous function to send to fsolve
x0=[0,0]; % Define starting point
xs=fsolve(fn,x0)   % Find the (x,y) location of the solution

% Make plot showing location of solution

x=0:0.05: 2; f1=cos(x);f2=x.^2;
plot(x,f1,'r',x,f2,'b'); hold on
plot(xs(1),xs(2),'bo','MarkerSize',12)
grid on
xlabel('x');ylabel('y')
axis([0 2 0 2]); axis square


function f=eq2(x) % This function defines the pair of equations

f(1)=x(2)-cos(x(1));
f(2)=x(2)-x(1)^2;

return
```

Figure C.4 shows the graphical output.

A slightly tidier version of the function above uses an anonymous function to replace the separate function that defines the system of equations. Its output is identical to Fig. C.4.

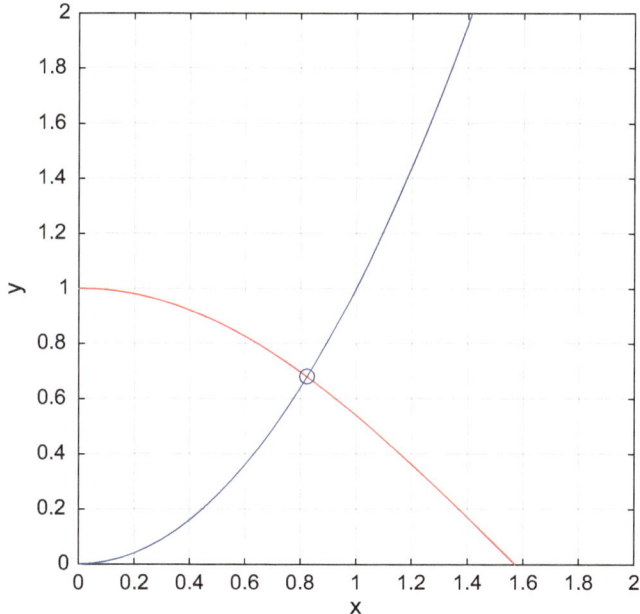

Fig. C.4 Solution of test equations using "fsolve"

```
function ex2_fsolve
%
% Demonstrate fsolve using an anonymous function
%
clc

fn=@(x) [x(2)-cos(x(1));x(2)-x(1)^2] % Anonymous function
x0=[0,0]; % Define starting point
xs=fsolve(fn,x0)   % Find the (x,y) location of the solution

% Make plot showing location of solution

x=0:0.05: 2; f1=cos(x);f2=x.^2;
plot(x,f1,'r',x,f2,'b'); hold on
plot(xs(1),xs(2),'bo','MarkerSize',12)
grid on
xlabel('x');ylabel('y')
axis([0 2 0 2]); axis square
```
.

Let's move on to finding the minimum of a two variable objective function using fsolve to solve the system $\nabla F(x) = 0$. Consider the butterfly function from Appendix B.

$$f(x,y) = \frac{x - y}{(x^2 + 5)(y^2 + 5)} + \frac{y^2}{20000} \qquad (C.5)$$

Conceptually, the simplest way to find the minimum using fsolve is to hard code the gradient, that is, to write it out explicitly as shown in the following M file.

```matlab
function ex3_fsolve_min
%
% Demonstrate fsolve to find minimum of Butterfly Function
%
clc

grad_f=@df % Make anonymous function from gradient
x0=[0,1]; % Define starting point
x_min=fsolve(grad_f,x0) % Find the location of the minimum

% Make contour plot showing location of solution
x=-4:0.01:4; [X,Y]=meshgrid(x);
obj=(X-Y)./((X.^2+5).*(Y.^2+5))+Y.^2/20000;
contour(X,Y,obj,'LevelStep',0.005)
grid on; axis square; hold on
xlabel('x');ylabel('y')
plot(x_min(1),x_min(2),'b+','MarkerSize',12)
text(-0.5,2,['x* = ' num2str(x_min(1))])
text(-0.5,1.5,['y* = ' num2str(x_min(2))])

return

function f=df(x)
%
%   This function defines two components of the gradient
%
f(1)=1/((x(1)^2+5)*(x(2)^2+5))-2*x(1)*(x(1)-x(2)) / ...
     ((x(1)^2+5)^2 * (x(2)^2+5));
f(2)=x(2)/10000-1/((x(1)^2+5)*(x(2)^2+5))- ...
     2*x(2)*(x(1)-x(2)) /  ((x(1)^2+5)^2 * (x(2)^2+5));

return
```

Figure C.5 shows the graphical output from this M file.

The previous example works well but requires writing out the objective function manually. A cleaner way to solve the problem is to ask MATLAB to find the gradient symbolically and then send the result to fsolve. The only disadvantage is that it requires some non-intuitive syntax to convert the gradient from a symbolic result to a numerical function. The following function shows one way to implement this approach.

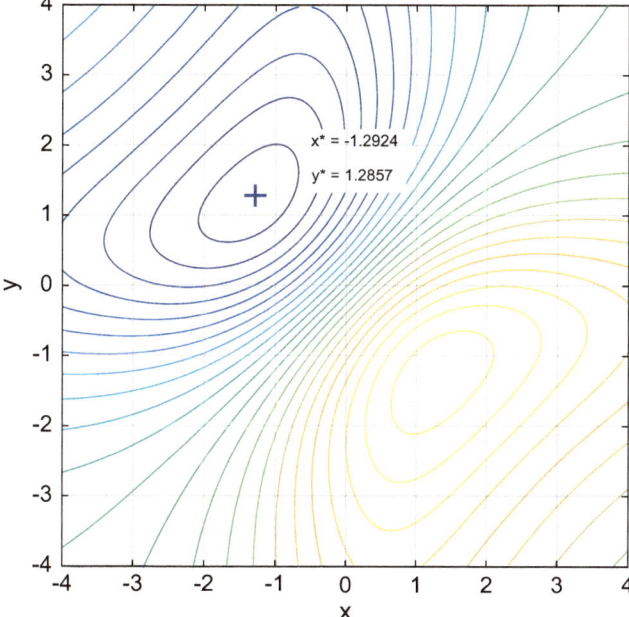

Fig. C.5 Graphical output of solution using "fsolve" to solve $\nabla F = 0$

```
function ex4_fsolve
%
%   Demonstrates fsolve to find minimum of Butterfly Function
%      Assistance with syntax from James Kristoff of MathWorks
%
clc

% Start Symbolic Calculations

syms x y
f=(x-y)/((x^2+5)*(y^2+5))+y^2/20000; % Define objective function
g=gradient(f,[x,y]); % Calculate gradient of objective function
ezcontour(f) % Draw contour plot of symbolic objective function
grid on
axis([-4 4 -4 4])
axis square
hold on

% Start Numerical Calculations

% Make numerical function
fn = {matlabFunction(g(1)), matlabFunction(g(2))};

% Make anonymous function from numerical function
fn1 = @(x)[fn{1}(x(1),x(2)), fn{2}(x(1),x(2))];

x0=[0,1];
x_min=fsolve(fn1,x0);
plot(x_min(1),x_min(2),'b+','MarkerSize',12)
text(-0.5,2,['x* = ' num2str(x_min(1))])
text(-0.5,1.5,['y* = ' num2str(x_min(2))])

return
```

Figure C.6 shows the graphical results from this M file. Note that there is more than one point in design space that approximately satisfies the requirement $\nabla F(x) = 0$. When the starting point was $x = 1$ and $y = 1$, fsolve returned the points $x* = 5.3302$ and $y* = 4.9481$.

Minimization Using fminsearch

Perhaps the simplest multidimensional minimization function in MATLAB is fminsearch. It uses an algorithm that doesn't require a derivatives, so it's very easy to implement. Omitting the graphical output results in a short M file.

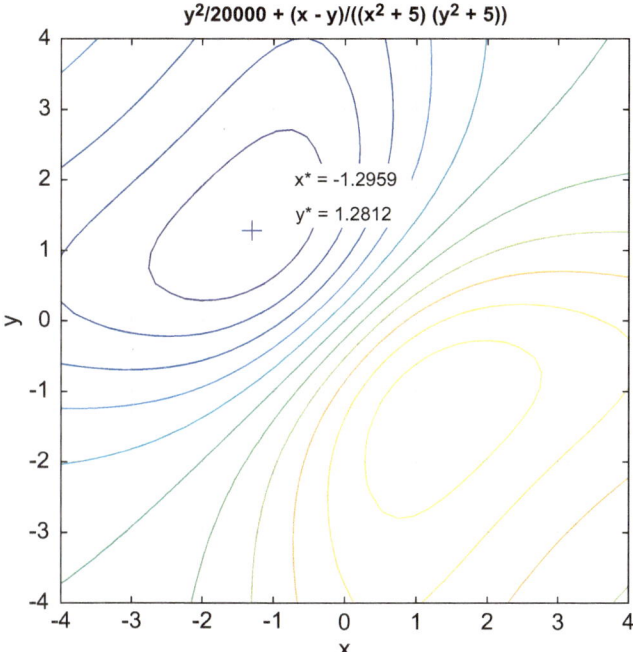

Fig. C.6 Graphical output using "fsolve" with symbolically calculated gradient

```
function c3_fminsearch
%
%   This function uses fminsearch to find the minimum
%   of the Butterfly Function
%
clc

% Define the objective function as an anonymous function
f=@(x) (x(1)-x(2))/((x(1)^2+5)*(x(2)^2+5))+x(2)^2/20000;

x0=[0,1]; % Define starting Point
x_min=fminsearch(f,x0); % Find minimum point using fminsearch

disp(['x* = ' num2str(x_min(1))])
disp(['y* = ' num2str(x_min(2))])

return
```

Text output:

$$x^* = -1.2959$$

$$y^* = 1.2813$$

The text output clearly shows that fminsearch found the essentially same point as did the previous examples.

A more sophisticated algorithm that can make use of derivatives is fminunc. In its simplest implementation, it can be treated just like fminsearch.

Unconstrained Minimization Using Fminunc

A more sophisticated unconstrained minimization function is fminunc. It uses several different algorithms and makes use of derivatives, though they are not required. Using it without derivatives is essentially a matter of replacing fminsearch with fminunc in the previous example.

```
function c4_fminunc
%
%   This function uses fminsearch to find the minimum
%   of the Butterfly Function
%
clc

% Define the objective function as an anonymous function
f=@(x) (x(1)-x(2))/((x(1)^2+5)*(x(2)^2+5))+x(2)^2/20000;

x0=[0,1]; % Define starting Point
x_min=fminunc(f,x0); % Find minimum point using fminsearch

disp(['x* = ' num2str(x_min(1))])
disp(['y* = ' num2str(x_min(2))])

return
```

The command window output of this M file is more involved since it includes a warning that no gradient was provided. The warning is informational and doesn't mean the result is incorrect.

```
Warning: Gradient must be provided for trust-region algorithm;
using quasi-newton algorithm instead.
> In fminunc (line 403)
  In c3_fminunc (line 12)
Local minimum found.

Optimization completed because the size of the gradient is less
than the default value of the function tolerance.

<stopping criteria details>


x* = -1.2959
y* = 1.2812
```

The next step is to add the gradient to the routine. This is done by creating a variable that defines the options for fminunc. It tells fminunc that the objective function gradient is available.

```
function c4_fminunc_grad
%
%   This function uses fminsearch to find the minimum
%   of the Butterfly Function
%
clc

x0=[0,1]; % Define starting Point

% Tell fminunc that the gradient is available
options=optimoptions('fminunc','GradObj','on'); % MATLAB R2015a

fn=@obj_grad; % fn contains both objective and gradient
x_min=fminunc(fn,x0,options); % Find minimum point using
fminsearch

disp(['x* = ' num2str(x_min(1))])
disp(['y* = ' num2str(x_min(2))])

return

function [f,g]=obj_grad(x)

% Define the objective function
f=(x(1)-x(2))/((x(1)^2+5)*(x(2)^2+5))+x(2)^2/20000;

%Define the gradient of the objective function
g(1) = 1/((x(1)^2+5)*(x(2)^2+5))-2*x(1)*(x(1)-x(2)) / ...
    ((x(1)^2+5)^2 * (x(2)^2+5));

g(2) = x(2)/10000-1/((x(1)^2+5)*(x(2)^2+5))- ...
    2*x(2)*(x(1)-x(2)) /  ((x(1)^2+5)^2 * (x(2)^2+5));

return
```

Command window output:

$$x* = -1.2932$$

$$y* = 1.2874$$

The point defined as the minimum varies slightly, as in some of the previous examples. This means that the tolerance for the objective function and the design variables have been reached. These tolerances can be modified using the optimoptions command.

Constrained Minimization Using Fmincon

The final step in our exploration of the basic MATLAB optimization functions is one that can address constrained minimization problems, fmincon. Consider the constrained optimization problem in Chap. 6.

$$\text{Minimize} \quad f(x,y) = \frac{x-y}{(x^2+5)(y^2+5)} + \frac{y^2}{20000} \qquad \text{(C.6)}$$

$$\text{Subject to} \quad g(x) = y + \sin(x) \le 0$$

The general form of this example follows the previous ones with two exceptions. The first is that fmincon allows for both equality and inequality constraints and inputs are required, even when one type of constraint is absent. In these cases, the blank input is []. In the example below, there are six blank inputs before the inequality constraint.

The other exception is a little less intuitive. The constraints are defined by a separate function that is required to return two entities. The first is the nonlinear inequality constraint, and the second is nonlinear equality constraint. If there is no equality constraint, as in this example, the equality constraint must be returned as a blank [].

```
function c5_fmincon
%
%   This function uses fmincon to find the minimum
%   of the Butterfly Function with constraint y+sin(x)<=0
%
clc

x0=[0,1]; % Define starting point in the feasible region
x_min=fmincon(@obj,x0,[],[],[],[],[],[],@con);

disp(['x* = ' num2str(x_min(1))])
disp(['y* = ' num2str(x_min(2))])

return

function f=obj(x)

% Define the objective function
f=(x(1)-x(2))/((x(1)^2+5)*(x(2)^2+5))+x(2)^2/20000;

return

function [g,ceq]=con(x)

% Define inequality constraint and bogus equality constraint
g=x(2)+ sin(x(1));
ceq=[];

return
```

The output to the command window is a little more involved, reflecting the wider range of options available:

```
Local minimum found that satisfies the constraints.

Optimization completed because the objective function is non-
decreasing in feasible directions, to within the default value of the
function tolerance, and constraints are satisfied to within the
default value of the constraint tolerance.

<stopping criteria details>


x* = -1.4762
y* = 0.99543
```

Index

© Springer International Publishing AG, part of Springer Nature 2018
M. French, *Fundamentals of Optimization*,
https://doi.org/10.1007/978-3-319-76192-3